Introduction to
Statistical and Machine Learning Methods for Data Science

Carlos Andre Reis Pinheiro
Mike Patetta

sas.com/books

Contents

About This Book

What Does This Book Cover?

This book gives an overview of the statistical and machine learning methods used in data science projects, with an emphasis on the applicability to business problem solving. No software is shown, and the mathematical details are kept to a minimum. The book describes the tasks associated with all stages of the analytical life cycle, including data preparation and data exploration, feature engineering and selection, analytical modeling considering supervised and unsupervised techniques, and model assessment and deployment. It describes the techniques and provides real-world case studies to exemplify the techniques. Readers will learn the most important techniques and methods related to data science and when to apply them for different business problems. The book provides a comprehensive overview about the statistical and machine learning techniques associated with data science initiatives and guides readers through the necessary steps to successfully deploy data science projects.

This book covers the most important data science skills, the types of different data science applications, the phases in the data science lifecycle, the techniques assigned to the data preparation steps for data science, some of the most common techniques associated to supervised machine learning models (linear and logistic regression, decision tree, forest, gradient boosting, neural networks, support vector machines, and factorization machines), advanced supervised modeling methods like ensemble models and two-stage models, the most important techniques associated to unsupervised machine learning models (clustering, association rules, sequence analysis, link analysis, path analysis, network analysis, and network optimization), the method and fits statistics to assess model results, different approaches to deploy analytical models in production, and the main topics related to the model operationalization process.

This book does not cover the techniques for data engineering in depth. It also does not provide any programming code for the supervised and unsupervised models, nor does it show in practice how to deploy models in production.

Is This Book for You?

The audience of this book is data scientists, data analysts, data engineers, business analysts, market analysts, or computer scientists. However, anyone who wants to learn more about data science skills could benefit from reading this book.

What Are the Prerequisites for This Book?

There are no prerequisites for this book.

We Want to Hear from You

SAS Press books are written *by* SAS Users *for* SAS Users. We welcome your participation in their development and your feedback on SAS Press books that you are using. Please visit sas.com/books to do the following:

- Sign up to review a book
- Recommend a topic
- Request information about how to become a SAS Press author
- Provide feedback on a book

About These Authors

Dr. Carlos Pinheiro is a Principal Data Scientist at SAS and a Visiting Professor at Data ScienceTech Institute in France. He has been working in analytics since 1996 for some of the largest telecommunications providers in Brazil in multiple roles from technical to executive. He worked as a Senior Data Scientist for EMC in Brazil on network analytics, optimization, and text analytics projects, and as a Lead Data Scientist for Teradata on machine learning projects. Dr. Pinheiro has a BSc in Applied Mathematics and Computer Science, an MSc in Computing, and a DSc in Engineering from the Federal University of Rio de Janeiro. Carlos has completed a series of postdoctoral research terms in different fields, including Dynamic Systems at IMPA, Brazil; Social Network Analysis at Dublin City University, Ireland; Transportation Systems at Université de Savoie, France; Dynamic Social Networks and Human Mobility at Katholieke Universiteit Leuven, Belgium; and Urban Mobility and Multi-modal Traffic at Fundação Getúlio Vargas, Brazil. He has published several papers in international journals and conferences, and he is author of Social Network Analysis in Telecommunications and Heuristics in Analytics: A Practical Perspective of What Influence Our Analytical World, both published by John Wiley Sons, Inc.

Michael Patetta has been a statistical instructor for SAS since 1994. He teaches a variety of courses including Supervised Machine Learning Procedures Using SAS® Viya® in SAS® Studio, Predictive Modeling Using Logistic Regression, Introduction to Data Science Statistical Methods, and Regression Methods Using SAS Viya. Before coming to SAS, Michael worked in the North Carolina State Health Department for 10 years as a health statistician and program manager. He has authored or co-authored 10 published papers since 1983. Michael has a BA from the University of Notre Dame and a MA from the University of North Carolina at Chapel Hill. In his spare time, he loves to hike in National Parks.

Learn more about these authors by visiting their author pages, where you can download free book excerpts, access example code and data, read the latest reviews, get updates, and more:

http://support.sas.com/pinheiro
http://support.sas.com/patetta

To Daniele, Lucas and Maitê.

Acknowledgments

I joined SAS on December 7th, 2015, but many people believed I had worked for SAS before. Not officially. But indeed, I have a long story with SAS.

I started using SAS in 2002 when I was working for Brasil Telecom, where I created a very active data mining group, developing supervised and unsupervised models across the entire corporation. In 2008, I moved to Dublin, Ireland, to perform a postdoc at Dublin City University. For two years I used SAS for social network analysis. I deployed SNA models at Eircom as a result of my research. After that, I spent six months at SAS Ireland using the brand new OPTGRAPH procedure. I developed some models to detect fraud in auto insurance and taxpayers.

In 2010, I returned to Brazil, and I had the opportunity to create an Analytics Lab at Oi. The Lab focused on developing innovative analytics for marketing, fraud, finance, collecting and engineering. SAS was a big sponsor/partner of it.

At the beginning of 2012, I worked for few months with SAS Turkey creating some network analysis projects for communications companies, and thereafter I moved to Annecy, France, to perform a postdoc at Université de Savoie, France. The research was focused on transportation systems, and I used SAS to develop network models. In 2013, I moved to Leuven, Belgium, to perform a postdoc at KU Leuven. The research was focused on dynamic network analysis, and I also used SAS for the model development. Back to Brazil in 2014, I worked as a data scientist for EMC^2 and Teradata, but most of the time I was still using SAS, sometimes with open-source packages. In 2014/2015, I performed a postdoc at Fundação Getúlio Vargas. The research was focused on human mobility and guess what, I used SAS.

Finally, thanks to Cat Truxillo, I found my place at SAS. I joined the Advanced Analytics group in Education. I have learned so much working at this group. It was a big challenge to keep up with such brilliant minds. I would like to thank each and every person in the Education group who has taught me over those years, but I would like to name a few of them specifically: Chris Daman, Robert Blanchard, Jeff Thompson, Terry Woodfield, and Chip Wells. To all of you, many thanks!

A special thanks to Jeff Thompson and Tarek Elnaccash for a relentless review. Both were instrumental in getting this book done.

Thanks to Suzanne Morgen for being an amazing editor and walking us through this process so smoothly.

Carlos Andre Reis Pinheiro

The idea for this book originated with Carlos Pinheiro. His experience as a data scientist has always impressed me, and this book highlights many of Carlos's success stories. Therefore, I would like to give thanks to Carlos for the inspiration for this book. I would like to thanks to the reviewers, Jeff Thompson, Tarek Elnaccash, and Cat Truxillo, for their diligent work to make the book technically accurate. Finally, I would like to give thanks to Suzanne Morgen, whose edits made the book flow as smoothly as possible.

Michael James Patetta

Foreword

The book you have open in front of you provides a taste of many data science techniques, interspersed with tales of real-world implementations and discoveries. The idea for this book originated when my team and I were designing the SAS Academy for Data Science. We designed a fairly ambitious training and certification program, assuming that people who enroll in the academy would have several years' experience working with data and analytics before they get started.

In 2015, the SAS Academy for Data Science was launched as a self-paced e-learning program. Designing the academy's curriculum required research into the state of data science, discussions with faculty training the next generation of data scientists, and shadowing consultants who bring the data to life for their clients. Those topics shift and evolve over time, and today, it is one of the top data science training programs in the world. The curriculum has been adopted by university graduate programs on every continent except Antarctica.

What we have found in practice, however, is that there is a considerably broader audience who want to enroll in the academy, including smart people who have experience in a different area, but do not have the benefit of several years' data analysis to guide their thinking of how they can apply analytics in their own fields.

For learners like these, where to begin? Carlos Pinheiro and Mike Patetta had the idea to create a short course that provides an overview of data science methods and lots of first-hand experiences as working data scientists.

Carlos Andre Reis de Pinheiro has written extensively in data science, including a Business Knowledge Series course (and later, his book) on Social Network Analysis. It was through this course that Carlos and I started working together. The first thing you notice about Carlos is that he is a born storyteller. The second thing you notice is that he loves soccer—I mean he really, really loves soccer. Over time I got to know more about this soccer-crazy professor who can keep everyone's attention with amazing stories from his data science research. Carlos has lived and worked in (at least) six different countries, and he is fluent in (at least) four languages. Here is a person with unstoppable curiosity and drive for growth. In 2016, he joined my colleagues and me in the Advanced Analytics Education department at SAS, where he has contributed his relentless hard work and ingenuity to solve business problems with data and analytics. Today he takes a direct, hands-on approach to showing companies what is possible with some data management elbow-grease, some well-trained models, and curiosity.

Mike Patetta has been a friend and colleague for over 20 years. In fact, he was the first person who interviewed me, in 1999, when I applied to work at SAS. Mike has a natural gift for educating others. He is someone who can dive into an unfamiliar topic in statistics and distill a shelf-full of books and journal articles down to a few learner-friendly hour-long lectures. The partnership between these two authors resulted in a course—and now, a book— that is rich with detailed information, written in an easy, comfortable style, with ample use cases from the authors' own experiences.

Data science is fun, or that's what recruiters would have you believe. Data science entails coaxing patterns, meanings, and insights from large and diverse volumes of messy data. In practice, that means spending more time than you might like on getting access to data, determining what is in a record, how records are represented in files, how the file is structured, and how to combine the information in a meaningful way with other files. That is, for many of us, most of the work a data scientist does. So where is the fun?

The reward of data science work comes when the data are organized, cleaned, and arranged for analysis. That first batch of visualizations, the feature engineering, the modeling—that is what makes data science such rewarding work. More than almost any other career, data scientists get to ask question after question, the answers leading to subsequent questions. From one day to another, your work can be completely different. You don't get to tell the data what to say—the data will speak to you, if you have the tools and curiosity to listen.

This book (and its accompanying course) provide a framework for doing project work, the analytics lifecycle. The analytics lifecycle acknowledges and addresses all members of the data science team—IT, computer engineers, statisticians, and executive stakeholders—and makes clear how the work and responsibilities are divided through the entire lifecycle of a data science project. The emphasis of this book is on *making sense*—of data, of models, and of results from deployed models. You might say that the ideal audience for this book is a Citizen Data Scientist (to use Gartner's term) or a statistical business analyst. This is not a book that teaches about writing scripts to pull hourly data, or how to stand up a virtual machine on a cloud hosting provider, or how to containerize your apps. This book alone will not turn you into a world-class data scientist. What it does, however, is take the loose ends of data science and organize them in a way that your project is manageable, and your results are explainable. And that, friends, is the core of any effective data and analytics professional.

Catherine Truxillo, PhD

Director, Advanced Analytics Education

SAS

February 2021

Chapter 1: Introduction to Data Science

Chapter Overview

This chapter introduces the main concepts of data science, a scientific field involving computer science, mathematics, statistics, domain knowledge, and communication. These are the main goals of this chapter:

- Explain the most important tasks and roles associated with the data science field.
- Explain how to apply data science to solve problems and improve business operations.
- Describe data science as a combination of different disciplines, such as computer science, mathematics and statistics, domain knowledge, and communications.
- Describe the skills related to mathematics and statistics and the role that they play in solving business problems through analytical modeling.
- Describe the skills related to computer science and the role that they play in solving business problems by supporting analytical models to be trained and deployed effectively.
- Describe the skills related to domain knowledge and the role that they play in solving business problems by adding value to data, models, and tactical and operational actions.
- Describe the skills related to communication and visualization and the role that they play in solving business problems by describing and explaining the solutions, model outcomes, and possible operational actions.

Data Science

Data science is not a single discipline. It comprises a series of different fields of expertise and skills, combining them to solve problems and to improve and optimize processes. Among several skills needed, the most important are mathematics and statistics, computer science, and domain knowledge.

Data scientists need mathematics and statistics to understand the data generated in the business scenario, to model this data to gain insights, or to classify or estimate future events. Mathematics and statistics are also needed to evaluate the models developed and assess how they fit to the problem and how they can be used to solve or improve a specific process.

The infographic in Figure 1.1 refers to three main areas: mathematics and statistics, computer science, and domain knowledge. In the next sections, we will discuss these areas more in depth as well as other important areas such as communication, visualization, and hard and soft skills.

Figure 1.1: Expertise Areas in Data Science

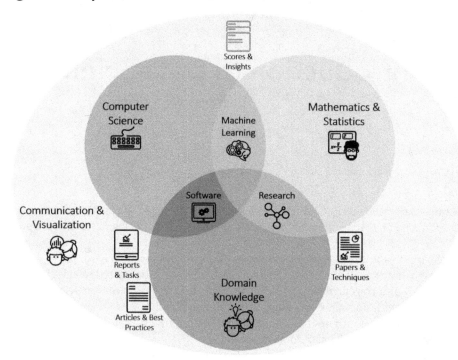

The intersection of each of the three main areas is also very important. *Machine learning* is the field intersecting mathematics, statistics, and computer science. Machine learning is a branch of artificial intelligence based on the idea that systems can learn from data, identify patterns, recognize behaviors, and make decisions with minimal human intervention. It is a method of data analysis that automates data preparation, feature engineering, model training, and eventually model deployment. Machine learning allows data scientists to implement very complex models, such as neural networks or support vector machines, and an ensemble of simple models like decision trees, gradient boosting, and random forests. These complex models can capture very unusual relationships between the inputs (independent variables) and the target (dependent variable).

The intersection of mathematics, statistics, and domain knowledge is the research field. Research skills enable data scientists to apply new techniques in model building. This combination allows the development of very complex models that are more accurate and less dependent on the functional form. Research skills can speed up the development process especially when fewer assumptions are needed about the distribution of the target and the relationship of independent and dependent variables.

Software skills in data science usually refer to the intersection of computer science and domain knowledge. Software skills such as familiarity of open-source languages and other world-class software languages help data scientists create new models. The combination of computer science skills, software skills, and domain knowledge can help data scientists solve the business problem or improve a specific business process.

Mathematics and Statistics

Data scientists need to have strong mathematics and statistics skills to understand the data available, prepare the data needed to train a model, deploy multiple approaches in training and validating the analytical model, assess the model's results, and finally explain and interpret the model's outcomes. For example, data scientists need to understand the problem, explain the variability of the target, and conduct controlled tests to evaluate the effect of the values of parameters on the variation of the target values.

Data scientists need mathematics and statistical skills to summarize data to describe past events (known as *descriptive statistics)*. These skills are needed to take the results of a sample and generalize them to a larger population (known as *inferential statistics)*. Data scientists also need these skills to fit models where the response variable is known, and based on that, train a model to classify, predict, or estimate future outcomes (known as *supervised modeling)*. These predictive modeling skills are some of the most widely used skills in data science.

Mathematics and statistics are needed when the business conditions don't require a specific event, and there is no past behavior to drive the training of a supervised model. The learning process is based on discovering previously unknown patterns in the data set (known as *unsupervised modeling)*. There is no target variable and the main goal is to raise some insights to help companies understand customers and business scenarios.

Data scientists need mathematics and statistics in the field of *optimization*. This refers to models aiming to find an optimal solution for a problem when constraints and resources exist. An objective function describes the possible solution, which involves the use of limited resources according to some constraints. Mathematics and statistics are also needed in the field of *forecasting* that is comprised of models to estimate future values in time series data. Based on past values over time, sales, or consumption, it is possible to estimate the future values according to the past behavior. Finally, mathematics and statistics are needed in the field of *econometrics* that applies statistical models to economic data, usually panel data or longitudinal data, to highlight empirical outcomes to economic relationships. These models are used to evaluate and develop econometric methods.

Mathematics and statistics are needed in the field of *text mining*. This is a very important field of analytics, particularly nowadays, because most of the data available is unstructured. Imagine all the social media applications, media content, books, articles, and news. There is a huge amount of information in unstructured, formatted data. Analyzing this type of data allows data scientists to infer correlations about topics, identify possible clusters of contents, search specific terms, and much more. Recognizing the sentiments of customers through text data on social media is a very hot topic called *sentiment analysis*.

Computer Science

The volume of available data today is unprecedented. And most important, the more information about a problem or scenario that is used, the more likely a good model is produced. Due to this data volume, data scientists do not develop models by hand. They need to have computer science skills to develop code, extract, prepare, transform, merge and store data, assess model results, and deploy models in production. All these steps are performed in digital environments. For example, with a tremendous increase in popularity, cloud-based computing is often used to capture data, create models, and deploy them into production environments.

At some point, data scientists need to know how to create and deploy models into the cloud and use containers or other technologies to allow them to port models to places where they are needed. Think about image recognition models using traffic cameras. It is not possible to capture the data and stream it from the camera to a central repository, train a model, and send it back to the camera to score an image. There are thousands of images being captured every second, and this data transfer would make the solution infeasible. The solution is to train the model based on a sample of data and export the model to the device itself, the camera. As the camera captures the images, the model scores and recognizes the image in real time. All these technologies are important to solve the problem. It is much more than just the analytical models, but it involves a series of processes to capture and process data, train models, generalize solutions, and deploy the results where they need to be. Image recognition models show the usefulness of containers, which packages up software code and all its dependencies so that the application runs quickly and reliably from one computing environment to another.

With today's challenges, data scientists need to have strong computer science skills to deploy the model in different environments, by using distinct frameworks, languages, and storage. Sometimes it is necessary to create programs to capture data or even to expose outcomes. Programming and scripting languages are very important to accomplish these steps. There are several packages that enable data scientists to train supervised and unsupervised models, create forecasting and optimization models, or perform text analytics. New machine learning solutions to develop very complex models are created and released frequently, and to be up to date with new technologies, data scientists need to understand and use these all these new solutions.

Software to collect, prepare, and cleanse data are also very important. Any model is just a mapping of the input data and the event to be analyzed, classified, predicted, or estimated. If the data is poor quality or has a lot of inconsistences, the model will map this, and the outcomes will be inaccurate. Huge amounts of data should be stored in efficient repositories. Databases are needed to accomplish that, and data scientists are required to understand how the databases work. To process massive amounts of data, distributed environments are required. Often, data scientists need to understand how these massive parallel processing engines work. In addition to databases, there are lots of new structures of repositories to perform analytics.

In summary, there are too many skills to be learned and mastered. Much more than one human being can handle. Therefore, in most of the cases, data scientists will need to partner with someone else to perform all the activities needed to create analytical models, such as data engineers, application developers, database administrators, infrastructure engineers, and domain knowledge experts.

Domain Knowledge

One of the most critical skills data scientists need is domain knowledge. Yes, human beings are still important. It is important to understand the problem and evaluate what is necessary to solve it. It is important to understand how to link the model results to a practical action. The analytical model gives a direction. How to use the results to really solve the problem, based on a series of information, policies, regulations, impacts, and so on, is the key factor to success. Another key factor in creating analytical models is awareness of the business problem and how it affects companies, communities, people, and governments. Knowing the business scenario helps data scientists create new input variables, transform or combine original ones, and select and discard important or useless information. This process called *feature engineering* includes creating new variables based on domain knowledge or based on machine learning. This is discussed in further chapters.

A good example of domain knowledge is in the telecommunications field. Data scientists need to know what type of data are available, how transaction systems are implemented, what billing system is used, and how data from the call centers are collected.

In addition to the domain knowledge, which allows data scientists to improve model development by incorporating business concepts, data scientists must be curious and try out multiple approaches to solve business problems. They need to be proactive in anticipating business issues and propose analytical solutions. There are situations where business issues are not clear, but improvements can be made. Data scientists need to be innovative to design and implement different approaches to solve problems. Creativity and innovation include combining distinct analytical models together to raise insights and to enrich the data analyses.

Collaboration is another key factor in data science. There are many fields of expertise, and it is almost impossible for a single person to master all of them. Collaboration allows data scientists to work alongside different professionals when seeking the best solution for a business problem. For example, if data scientists are developing models to predict loan defaults, then they probably need to partner with someone from the finance department to help them understand how the company charges the customers, what time frame is used to define a default, and how the company deals with defaults. All these policies should be considered when developing data analyses and training analytical models. The data associated with this problem most likely sit in different transactional systems with distinct infrastructures. Data scientists need to work with data engineers, software developers, and informational technology operation personnel to gather all the data sources in an effective way. Once the model is done and deployment is required, data scientists need to work with application developers to make the model's outcomes available to the organization in the way it is expected. The combination of all these skills creates an effective analytical framework to approach business problems in terms of data analysis, model development, and model deployment.

Communication and Visualization

One more key skill is essential to analyze and disseminate the results achieved by data science. At the end of the process, data scientists need to communicate the results. This communication can involve visualizations to explain and interpret the models. A picture is worth a thousand words. Results can be used to create marketing campaigns, offer insights into customer behavior, lead to business decisions and actions, improve processes, avoid fraud, and reduce risk, among many others.

Once the model's results are created, data scientists communicate how the results can be used to improve the operational process with the business side of the company. It is important to provide insights to the decision makers so that they can better address the business problems for which the model was developed. Every piece of the model's results needs to be assigned to a possible business action. Business departments must understand possible solutions in terms of the model's outcomes and data scientists can fill that gap.

Data scientists use visual presentation expertise and story-telling capabilities to create an exciting and appealing story about how the model's results can be applied to business problems. Data analysis and data visualization sometime suffice. Analyzing the data can help data scientists to understand the problem and the possible solutions but also help to drive straightforward solutions with dashboards and advanced reports. In telecommunications, for example, a drop in services consumption can be associated with an engineering problem rather than a churn behavior. In this case, a deep data analysis can drive the solution rather than a model development to predict churn. This could be a very isolated problem that does not demand a model but instead a very specific business action.

Hard and Soft Skills

Hard skills include mathematics, statistics, computer science, data analysis, programming, etc. On the other side, there are a lot of soft skills essential to performing data science tasks such as problem-solving, communication, curiosity, innovation, storytelling, and so on. It is very hard to find people with both skill sets. Many job search sites point out that there is a reasonable increase in demand for data scientists every year. With substantial inexpensive data storage and increasingly stronger computational power, data scientists have more capacity to fit models that influence business decisions and change the course of tactical and strategical actions. As companies become more data driven, data scientists become more valuable. There is a clear trend that every piece of the business is becoming driven by data analysis and analytical models.

To be effective and valuable in this new evolving scenario, data scientists must have both hard and soft skills. Again, it is quite difficult to find professionals with both hard and soft skills, so collaboration as a team is a very tangible solution. It is critical that data scientists partner with business departments to combine hard and soft skills in seeking the best analytical solution possible.

For example, in fraud detection, it is almost mandatory that data scientists collaborate with the fraud analysts and investigators to get their perspective and knowledge in business scenarios where fraud is most prevalent. In this way, they can derive analytical solutions that address feasible solutions in production, usually in a transactional and near real-time perspective.

Data Science Applications

It is difficult to imagine a company or even a segment of industry that cannot benefit from data science and advanced analytics. Today's market demands that all companies, private or public, be more efficient and accurate in their tactical and operational actions. Analytics can help organizations drive business actions based on data facts, instead of guesses, as shown in Figure 1.2.

Another key factor in data science is that all the techniques data scientists use in one industry can be easily applied to another. Business problems, even in different industries, can be remarkably similar. Insolvency is a problem in many industries: telecommunications, banking, and retail, for example. The way banks handle

Figure 1.2: Industries That Can Benefit from Data Science Projects

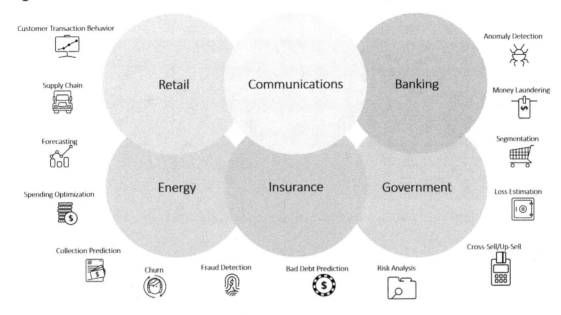

insolvency can help retail and telecom companies improve their process and be more effective in their business. How telecommunications companies handle transactional fraud can help banks improve their process to detect and react to fraudulent transactions in credit card accounts. Data analysis and analytical insights need to replace guessing and assumptions about market, customers, and business scenarios.

Collaboration among data scientists working in different industries is valuable and increases the spectrum of viable solutions. Industries can learn from each other and improve their process to identify possible analytical solutions and deploy practical business actions. This knowledge transferring from different domain areas, even between distinct industries, is beneficial for all sides involved. Several current business issues include fraud detection, churn analysis, bad debt, loss estimation, cross-sell/up-sell, risk analysis, segmentation, collecting, optimization, forecasting, supply chain, and anomaly detection, among many others.

Data Science Lifecycle and the Maturity Framework

The analytical lifecycle, or the data science process flow, comprises a few steps. However, it is important to spend some time on them and make sure all of them are performed properly.

Understand the Question

The first step in the modeling process as shown in Figure 1.3 is to understand the question. Data scientists need to understand what they are trying to solve with the model that they are about to develop. To perform this step, work closely with the business department to verify if a model is

Figure 1.3: Data Science Projects Lifecycle

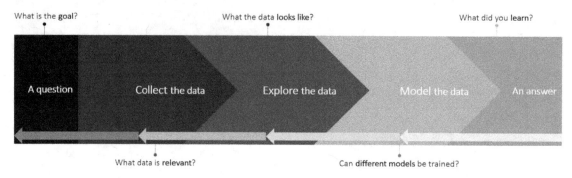

appropriate, if it is feasible, if there is enough data to use, and what practical actions are planned to be deployed based on the model's outcomes.

Some of the questions to ask during this phase are:

- What is the goal of the project? Do we want to predict some future event or classify, estimate, or forecast? Do we want to optimize a specific process or generate insights about customer behavior? Do we want to create groups or segments or produce a path analysis, sequence analysis, network analysis, and so on?
- What is the specific objective of the model? Is it a supervised or unsupervised model? Is the data structured or unstructured?
- Is there enough data to address this problem?
- What actions are planned based on the model's outcomes?

For example, in a churn model, you need to produce an actionable definition of churn. Is the model a classification (yes or no), and can a campaign be triggered based on the likelihood of churn? Multiple approaches or customer offerings can be assigned to the probability of customer churn. The response can be captured to produce feedback to the model and to improve the performance in subsequent actions.

Collect the Data

The second step in the data science process flow is to collect the data. Most likely, this phase requires multiple people and different skills. Database management, repositories, programming, data quality packages, data integration, and many other technologies might be required to accomplish this step properly.

Some of the pertinent questions during this phase are:

- What data is relevant?
- How many data sources are involved?
- Where do the data sources reside?
- Is the access to the data readily available?

- Are there any privacy issues?
- Are the data available when the model is deployed in production?

For example, some variables, such as gender and income, might not be available to use even though those predictors might be associated with the outcome. Furthermore, even if the data are available when you develop the model, are the data available when the model is in production, such as scoring a business transaction for fraud? Is it possible to access all the data used during the model training when the model is used for scoring? Are there any data privacy regulations? This is a problem because scored observations with missing values for the predictor variables in the scoring model have missing predicted outcomes.

Explore the Data

The third step is to explore the data and evaluate the quality and appropriateness of the information available. This step involves a lot of work with the data. Data analysis, cardinality analysis, data distribution, multivariate analysis, and some data quality analyses–all these tasks are important to verify if all the data needed to develop the model are available, and if they are available in the correct format. For example, in data warehouses, data marts, or data lakes, customer data is stored in multiple occurrences over time, which means that there are multiple records of the same customers in the data set. For analytical models, each customer must be a unique observation in the data set. Therefore, all historical information should be transposed from rows to columns in the analytical table.

Some of the questions for this phase are:

- What anomalies or patterns are noticeable in the data sets?
- Are there too many variables to create the model?
- Are there too few variables to create the model?
- Are data transformations required to adjust the input data for the model training, like imputation, replacement, transformation, and so on?
- Are tasks assigned to create new inputs?
- Are tasks assigned to reduce the number of inputs?

In some projects, data scientists might have thousands of input variables, which is far too many to model in an appropriate way. A variable selection approach should be used to select the relevant features. When there are too few variables to create the model, the data scientist needs to create model predictors from the original input set. Data scientists might also have several input variables with missing values that need to be replaced with reasonable values. Some models require this step, some do not. But even the models that do not require this step might benefit from an imputation process. Sometimes an important input is skewed, and the distribution needs to be adjusted. All these steps can affect the model's performance and accuracy at the end of the process.

Model the Data

The fourth step is the analytical model development itself. Some say that this is the most important part, or at least, where data scientists have more fun. Here they will use their creativity and innovation skills to try out multiple analytical approaches to solve the business problem. As stated before, data science is a mix of science and art. This step is the time when data scientists apply both the science behind all algorithms and the art behind all analytical approaches.

Some questions to consider at this phase are:

- Which model had the highest predictive accuracy?
- Which model best generalizes to new data?
- Is it possible to validate the model? Is it possible to test the model? Is it possible to honestly test the models on new data?
- Which model is the most interpretable?
- Which model best explains the correlation between the input variables and the target? Which one best describes the effects of the predictors to the estimation?
- Which model best addresses the business goal?

This is the data scientists' playground, where they use different algorithms, techniques, and distinct analytical approaches! Yes, a lot of the modeling process involves simply trying new algorithms and evaluating the results. Data science differs from some exact sciences, like math and physics, where based on a robust equation and inputs, it is possible to predict the output. In data science, the set of inputs might be known, but the exact subset of predictors is still unknown until the end of the model training. The equation is created during the model training according to the input data. Then the results are revealed. Any change in the input data set implies a change in the output. Therefore, data science is very much tied to the statistical and mathematical algorithms. However, all the rest is art. Furthermore, many models are not robust as they should be. Some models or algorithms are very unstable, which means every training data set might represent a different result.

Maybe this is the fun part. In this phase, data scientists try to fit the model on a portion of the data and evaluate the model's performance on another part of the data. The first portion is the training set. The second one is the validation set. Sometimes there is a third portion called the test set. It should be noted that sometimes the best model, depending on the business goal, is the most interpretable and simplest model, rather than the one with the highest predictive accuracy. It depends on the business goal, the practical action, and if there is any regulation in the industry.

In summary, this fourth step includes the following tasks:

- Train different models (algorithms/techniques/analytical approaches).
- Validate all trained models based on a different data set.
- If possible, test all models trained on a different data set (different from the one used during the validation).
- Assess all models' outcomes and evaluate the results based on the business goals.
- Select the best model based on the business requirements.
- Deploy and score the best model to support the business action required.

Perhaps one of the most difficult phases of the analytical process is to analyze the results and evaluate how the model's outcomes can support the business action required. This step is somehow related to the previous one, when data scientists train multiple models and assess the results. At this phase, domain knowledge and communication skills play a key role.

Provide an Answer

The fifth and last step is to provide answers to the original questions, the ones raised and validated during the first step. Some pertinent questions are:

- What lessons were learned from the trained models?
- How do the trained models answer the original questions?
- How do the trained models tell a story and support business decisions?
- How can the trained models be used and deployed in production in the appropriate time frame to support the required business actions?

For example, can the trained model support a campaign that targets customers with the highest probabilities of churn and offer them incentives to keep them using and consuming the company's products and services? Can the trained model support a fraud detection process in real time to identify possible fraudulent business transactions? The model's results might be very accurate, but to benefit the organization, the model should be deployed in an appropriate time frame. For example, in cybersecurity, if the model does not generate real-time alerts in a way that fraud analysts can take immediate actions, then the model might be useless, since digital attacks must be identified within seconds, not weeks or months.

Once an answer is provided, it might generate more questions regarding the business problem. Therefore, the data science lifecycle is cyclical as the process is repeated until the business problem is solved.

The entire analytical process and the data science approach can be viewed as a dynamically evolving flow as shown in Figure 1.4. In data science, the more complex the analytical task, the more value added to the business. For example, a simple query report can add value to the business by simply illustrating the relationships in the data, showing what happened in the past. It is very much descriptive in the sense that nothing can be done to change that historical event. However, awareness is the first step to understand the business problem and aim for an analytical solution.

Data exploration analyses can add further value to the business with more complex queries to the data. Multi-dimensional queries can help business analysts to not only understand what happened, but why it happened in that way. Analyzing the historical data under multiple dimensions at the same time can answer many questions about the business, the market, and the scenarios. Data mining, analytics, or data science, regardless of the name, it is a further step to gain knowledge about the business. Some analytical models explain what is going on right now. Unsupervised models such as clustering, segmentation, association analysis, path analysis, and link analysis help business analysts understand what exactly happens in a very short time frame and allow companies to deploy business actions to take advantage of this knowledge. Furthermore, supervised models can learn from past events and predict and estimate future occurrences. Data science in this phase is basically trying to know what will happen in the future. This is very similar to econometric and forecast models trying to foresee what will happen soon with a business event.

The final stage in the evolving analytical process is optimization. Optimization algorithms add more business value by showing what specific offer to make each customer. Optimization models consider an objective function (what to solve), a limited set of resources (how to solve), and a set of constraints

Figure 1.4: Model's Evolution in Advanced Analytics

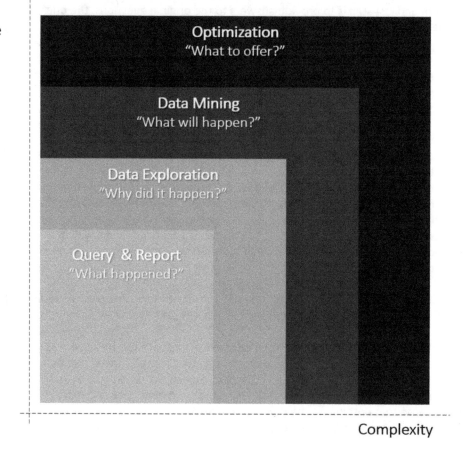

(solve it at what price). An organization might have several models in production to classify customers in multiple aspects, for example, the likelihood to make churn, the probability to not pay, etc. A combination of all these scores can be used to optimize campaigns and offerings. For example, the churn model predicts who is the customer most likely to make churn. However, not all customers have the same value to the business. Some might be insolvent. Some might not generate any profit. Some can be very valuable. The optimization process shows what incentives to offer to certain customers to maximize the profit in a specific campaign.

Advanced Analytics in Data Science

Data science and advanced analytics comprise more than simply statistical analysis and mathematical models. The field encompasses machine learning, forecasting, text analytics, and optimization. Data scientists must use all these techniques to solve business problems. In several business scenarios, a combination of some of these models are required to propose a feasible solution to a specific problem.

There are basically two types of machine learning models: *supervised learning* (when the response variable (also known as the *target*) is known and used in the model) and *unsupervised learning* (when the target is unknown or not used in the model). The *input variables* (also called features in the machine learning field or the independent attributes in the statistical field) contain information about the customers, who they are, how they consume the product or service, how they pay for it, for how long they are customers, from where they came from, where they went to, among many other descriptive information.

The target is the business event of interest, for example, when a customer makes churn, purchases a product, makes a payment, or simply uses a credit card or makes a phone call. This event is called a target because this is the event the model will try to predict, classify, or estimate. This is what the company wants to know. (A target is also called a *label* in the machine learning field or *dependent attribute* in the statistical field.) Unsupervised models do not require the target. These models are used to generate insights about the data, or market, or customers, to evaluate possible trends or to better understand some specific business scenarios. These models do not aim to classify, predict, or estimate a business event in the future.

As shown in Table 1.1, regression, decision tree, random forest, gradient boosting, neural network, and support vector machine are examples of supervised models. Clustering, association rules, sequence association rules, path analysis, and link analysis are examples of unsupervised models. There is a variation of these types of models called semi-supervised models. *Semi-supervised models* involve a small amount of data where the target is known and a large amount of data where the target is unknown. There are also models associated with reinforcement learning, where the algorithm is trained by using a system to reward the step when the model goes in the right direction and to punish the step when the model goes in the wrong direction. Semi-supervised models are becoming more prevalent and are often implemented in artificial intelligence applications. For example, reinforcement learning can be used to train a model to learn and take actions in self-driving cars. During the training, if the car drives safely in the road, the learning step is rewarded because it is going in the right direction. On the other hand, if the car drives off the road, the learning step is punished because the training is going in the wrong direction.

As statistical models try to approximate reality through mathematical formalized methods making predictions about future events, machine learning automates some of the most important steps in analytical models, the learning process. Machine learning models automatically improve the learning steps based on the input data and the objective function.

Table 1.1: Machine Learning Models

Supervised Models	Unsupervised Models
Regression	Clustering
Decision Tree	Association Rules
Random Forest	Sequence Association Rules
Gradient Boosting	Path Analysis
Neural Networks	Link Analysis
Support Vector Machine	

Data scientists should be able to build statistical and machine learning flows to access the available data, prepare the data for supervised and unsupervised modeling, fit different types of analytical models to solve the business problems, evaluate models according to the business requirements, and deploy the *champion model* (the chosen model based on some criteria) and *challenger models* (models that are trained differently such as using different algorithms) in production to support the planned business actions. All analytical models are trained based on data sets that, considering a specific time frame, describe the market scenario and the business problem. The models learn from the input variables and create a generalized map to the target, establishing relationships between the input variables and the target. This map describes the past behavior associated with the target in a specific point in time. As time goes by, customer behavior might change and thus the data that describes that pattern. When this happens, the current model in production drifts because it is based on a past customer behavior that no longer exists. That model needs to be retrained or a new model needs to be developed. This is a very important cycle in analytics, and the field of machine learning can contribute greatly with models that automatically retrain or learn from experience.

There is a great debate about when and how to use machine learning models and statistical models. Usually, machine learning models can be more accurate and perform better in production to support business actions. As a caveat, most of the machine learning models are not easy to interpret or explain. On the other hand, statistical models can generalize better estimations for future events. They are often simpler and easier to interpret and explain. In some industries, usually the strictest regulated ones, an interpretable model is mandatory. Statistical models also make the inputs and their effects on the prediction easier to explain, allowing the business departments to design campaigns and promotions more geared to the customers' behaviors.

Statistical analysis is the science of collecting, exploring, and presenting large amounts of data to discover underlying patterns, behaviors, and trends. Organizations use statistics and data analysis every day to make informed business decisions. As more data are collected every day and the infrastructure to store and process all this data gets cheaper, more data analyses are performed to drive business decisions. Statistical analysis includes *descriptive statistics*, where models summarize the available data to describe past events or previous scenarios. Another statistical analysis field is *inferential statistics*, where models take the results of a sample and generalize and extrapolate them to a larger population. Another field in statistical analysis is *predictive modeling*, in which models provide an estimate about the likelihood of a future outcome. This outcome can be a binary target, a multinomial target, or a continuous target. A binary target is assigned to a classification model, like yes or no. A multinomial target is assigned to a type of classification, but for multiple classes, like high, medium, and low. A continuous target is assigned to an estimation, where the event can be any continuous values, like the loss in a fraud event or the amount of purchase. For example, a credit score model can be used to determine whether a customer will make a future payment on time or not. The credit score model can also classify the range of the risk, such as high risk to default, medium risk to default or low risk to default. The credit score model can finally estimate the value associated to the default.

Finally, there is the field of *prescriptive statistics*, where models quantify the effect of future decisions. This area simulates and evaluates several possible courses of action and allows companies to assess different possible outcomes based on these actions. It is like a what-if type of analysis.

Forecasting describes an observed time series to understand the underlying causes of changes and to predict future values. It involves assumptions about the form of the data and decomposes time series data into multiple components. *Auto-regressive integrated moving average (ARIMA) models* are forecasting models where the predictions are based on a linear combination of past values, past errors, and current and past values of other time series. Another type of forecasting model is the *causal model*, which forecasts time series data that are influenced by causal factors such as calendar events to describe possible seasonality. Finally, there are modern and complex forecasting models that incorporate time series data whose level, trend, or seasonal components vary with time. They might include hierarchical segments of time series and recurrent neural networks to account for stationary and nonstationary data.

Text analytics is a field associated with uncovering insights from text data, usually combining the power of natural language processing, machine learning, and linguistic rules. Data scientists can use text analytics to analyze unstructured text, extract relevant information, and transform it into useful business intelligence. For example, data scientists can use information retrieval to find topics of an unstructured document, such as using a search engine to find specific information. Sentiment analysis is another field in text analytics and very useful in business. *Sentiment analysis* determines levels of agreement from unstructured data associating the overall information as a positive, negative, or neutral sentiment. Data scientists also use text analytics for topics discovery and clustering, where topics or clusters are revealed from various text documents based on the similarity that they have between them.

Finally, *text categorization* is a technique where a text analytics model labels natural language texts with relevant categories from a predefined set. Domain experts and linguistics interact in this field to create and evaluate the categories.

Survival analysis is a class of statistical methods for which the outcome variable of interest is the time until an event occurs. Time is measured from when an individual first becomes a customer until the event occurs or until the end of the observation interval (the individual then becomes censored). In survival analysis, the basis of the analysis is tenure, or the time at risk for the event. Therefore, it is not just whether the event occurred, but when it occurred.

The goal in survival analysis is to model the distribution of time until an event. The job of the data scientist is to identify the variables that are associated with the time until an event and to predict the time until an event for new or existing customers. Survival analysis can be used in customer retention applications where the outcome is the time until the cancellation of all products and services. Another example is credit risk management applications where the outcome is the time until a loan defaults.

The last topic in the advanced analytics framework is *optimization*. Mathematical optimization is a major component in operations research, industrial engineering, and management science. An optimization model searches for an optimal solution, which considers a set of pre-determined constraints and a limited set of resources.

For example, in production planning, optimization models can determine the best mixes of products to be produced to achieve the highest profit. In pricing decisions, optimization models can determine the optimal price for products based on costs, demands and competitive price information. Finally, in promotional marketing, optimization models can determine the best combination of promotional offers, delivery channels, time for campaigns, and the best set of customers to be contacted to maximize the return of the marketing investment.

Another important area of optimization is network analysis and network optimization, which involve analysis of networks (nodes and links) and analysis of network flow. For example, data scientists can use network optimization to study traffic flows where the nodes are cities, and the thickness of the links indicate traffic flow. This information can be used to plan for road widening projects and to construct new roads.

Data Science Practical Examples

Data science and advanced analytics can help organizations in solving business problems, addressing business challenges, and monetizing information and knowledge. Areas like customer experience, revenue optimization, network analytics, and data monetization are just few examples.

Customer Experience

Customer experience and engagement management allow organizations to gain a deeper understanding of the customer experience and how customers respond to the company's stimulus. Some examples include:

- Enhanced Customer Experience – offer more personalized and relevant customer marketing promotions.
- Insolvency and Default Prediction – monitor expenses, bills, and usage over time within safe ranges allowing customers to be on time in their payment events.
- Churn Prediction and Prevention – forecast customer issues and prevent them from happening. For example, if a utility company forecasts an outage and sends a text to the customers affected by the outage, this might improve the customer experience.
- Next Best Offer – operationalize customer insights by using structured and unstructured data. For example, modeling the customer's purchasing history can predict what the next item is to offer the customer.
- Target Social Influencers – identify and track relationships between customers and target those with the most influence. These influencers within their networks can help companies to avoid churn and increase product adoption.

Revenue Optimization

Revenue optimization includes better forecasts to allow organizations to make better business decisions. Some examples include:

- Product Bundles and Marketing Campaigns – improve business decision-making processes related to product bundles and marketing campaigns.
- Revenue Leakage – identify situations leading to revenue leakage, whether due to billing and collections, network, or fraud issues. For example, a collection agency might be interested in who is more likely to pay their debts rather than who owes the most money.
- Personalize products and services – personalize packages, bundles, products, and services according to customers' usages over time and allow them to pay for what they consume.
- Rate Plans and Bundles – use advanced analytics to create new and more effective rate plans and bundles.

Network Analytics

Network analytics is very common in communication and utility industries, but it can be applied in almost any type of industry, in the sense that most complex problems can be viewed as a network problem. Network analytics aims to improve network performance. Network performance can refer to a communications network, energy network, computer systems network, political network, or supply chain network, among many others. Some examples include:

- Network Capacity Planning – use statistical forecasting and detailed network or supply chain data to accurately plan capacity. Use forecasting on ATM machines to make sure there is enough cash to meet customer demands or network analytics to make sure customers have a good mobile signal wherever they go.
- Service Assurance and Optimization – use network analytics to prevent network or supply chain problems before they happen, either in communications, utilities, or retail.
- Optimize supply chain – identify the best routes (cheapest, fastest, or shortest) to deliver goods across geographic locations to keep customers consuming products and services in a continuous fashion.
- Unstructured Data – use unstructured data for deeper customer and service performance insights. For example, optimizing call center staffing and identifying operational changes could lower the cost to serve the customers while improving service quality.

Data Monetization

Information is the new gold. Data science can be used for organizations to monetize the most important asset that they have, which is data. *Data monetization* refers to the act of generating measurable economic benefits from available data sources. Companies in different industries have very sensitive and important data about customers or people in general. Think about telecommunications companies that have precise information about where people go at any point in time. Where people are in a point in time can be easily monetized. Data scientists can use the results of mobile apps that track what customers do, when, and with who. They can also use the results of web search engines that know exactly what customers are looking for in a point in time. Some examples of data monetization include:

- Location-Based Marketing – develop specialized offers and promotions that are delivered to targeted customers via their mobile devices. For example, if you enter a specific area in a city, you might get a coupon for companies located in that area.
- Micro Segmentation – create highly detailed customer segments that can be used to send very specific campaigns, promotions, and offerings over time.
- Third-party Partnerships – partner with different companies to combine customer data to enrich the information used to create analytical models and data analyses about business actions.
- Real-Time Data Analysis – analyze real-time data streams from different types of transactions to keep customers consuming or using products and services with no outages or intermittent breaks.

In conclusion, data scientists can develop and deploy a set of techniques and algorithms to address business problems. Today, companies are dealing with information that comes in varieties and volumes never encountered before. As data scientists increase their skills in areas such as machine learning, statistical analysis, forecasting, text analytics, and optimization, their value to the company will increase over time.

Summary

This chapter introduced some of the most important concepts regarding analytical models for data science. It considers the aspects of hard skills, like computer science, mathematics and statistics, and soft skills, such as domain knowledge and communication. This chapter briefly presented tasks associated with the analytical lifecycle and how some of those data science analytical models can be applied to solve business problems. Different hard and soft skills are crucial in data science projects when trying to solve business problems. Each one of these skills contributes to various phases of the data science project, from preparing and exploring the data, training the analytical models, evaluating model results, and probably the most important phases, deploying the analytical models into production and communicating the model's results to the business departments. This communication is important to help the business areas to define the campaigns, the offerings, and all the strategies to approach customers.

Now that you have an overview of the data science field, the next chapter will discuss data preparation.

Additional Reading

1. Davenport, T.H. and Patil, D.J., "Data Scientist: The Sexiest Job of the 21st Century," *Harvard Business Review*, October 2012.
2. Flowers, A., "Data Scientist: A Hot Job that Pays Well," Indeed Hiring Lab, January 2019.

Chapter 2: Data Exploration and Preparation

Chapter Overview

This chapter introduces the main concepts of data exploration, shows how data exploration is a key stage for any data science project, and shows how it can address many business issues as a result of reviewing and exploring the information needed to solve a business problem. In addition to data exploration, this chapter introduces the many tasks associated with data preparation, particularly focusing on pre-steps for analytical modeling. During the data preparation process, many data issues can arise. You might need to fix data quality issues, transform the raw information, and impute and replace original values to fit analytical models. This is an essential step in any data science project.

This chapter also introduces concepts about merging data from different sources to create a more comprehensive view about the business problems, considering customer behavior, product and service transactions, marketing scenarios, profiles, and so on. Usually, different information resides in multiple systems, and merging and matching them together is a great challenge in data preparation. You will also learn how these tasks can substantially improve results.

The main goals of this chapter are:

- Explain how to review data sets to determine whether the information needed to address the business problem is included and properly formatted to be used in analytical models to solve the problem.
- Explain how to perform distribution and cardinality analyses to determine how the data is presented and what needs to be done to correct the data for data science projects and analytical models.
- Explain how to handle non-linearities, sparseness, and outliers in the data used to train analytical models to solve business problems.
- Explain how to prepare and use unstructured data to improve analytical models in solving business problems. Unstructured data can be used directly in solving the business problems or can be used as new features in analytical models when solving business problems.
- Explain different methods of sampling data. Describe when each method of sampling is required and the benefits to the analytical modeling stage.
- Explain the method of partitioning data, the distinct types of partitioning, and the reason for partitioning data to generalize analytical models for production.
- Explain how to transform the original input data to improve the results in analytical modeling.

- Explain how to create new features from the original input data to improve the results in analytical modeling.
- Explain how to select a model that generalizes well to new data.

Introduction to Data Exploration

One of the first steps in the analytical modeling process is data exploration. This step is where data scientists get to know the data that they will use to solve any business problem. There are many tasks that data scientists go through not just to understand the data that describes the business problem that they are trying to solve, but also to prepare that data to solve those problems in terms of analytical models.

Tasks associated with data exploration include managing the following:

- Nonlinearity
- High cardinality
- Unstructured data
- Sparse data
- Outliers
- Mis-scaled input variables

Nonlinearity

Data exploration can illustrate nonlinear associations between the predictor variable and the outcome. An example of a nonlinear association is a quadratic relationship between the predictor variable and the outcome. In this situation, the data scientists must modify the model to consider the quadratic association.

Nonlinearity demands that data scientists explore different models to account for the more complex relationships between the predictor variables and the target. Universal approximator models such as tree-based models and neural networks can be used here to mitigate these complex relationships.

High Cardinality

Some variables might have high cardinality where categorical variables have numerous levels. Variables such as ZIP codes or product codes might have high cardinality. The high cardinality can be very problematic in analytical models like regression or neural networks because for each level it creates a parameter estimate or a weighted parameter, respectively. These levels essentially create models that are much more complex at the end of the process.

Suppose you are trying to find a linear association between salary (the target) and years of experience. You have two parameters, an intercept parameter, and the parameter for years of experience. If age is added, you now have three parameters. However, if educational background is added to the model as a categorical variable (less than high school, high school degree, and so on) now you might add six new parameters. Consequently, one categorical predictor variable can add much complexity to the analytical model. Therefore, cardinality is something that all data scientists need to consider.

Unstructured Data

Using unstructured data such as textual data, network data, images, audio, and video might help explain the business problem or simply hold a greater predictive power in relation to the outcome data scientists are modeling. For example, in a churn analysis, examining call center data and identifying unhappy customers might lead to good predictors for churn. This data is usually in textual format or audio format. How customers relate to each other and how products and services are related to the customers can hold a great predictive value because they describe consuming and usage behavior and its correlation to some specific business events, such as churn, fraud, or purchase. These types of data should be prepared and used in analytical models either to solve the business problems directly or to add new features for improving model performance and interpretability.

Sparse Data

Data scientists also encounter sparse data where there are very few events in the data and most of the occurrences are missing or incomplete. An example is a rating system for an online retail company. Not all customers who purchase a company's products rate the product. Conversely, not all products are rated by the customers. When combining all products and all customer ratings, the matrix generated is very sparse. It creates a serious problem for most models in capturing the relationship between the inputs and the target. One problem is that there is not enough data to estimate the interactions between the customers and the products.

There is a model tailored for this type of problem. Factorization machine models circumvent the problem of sparse matrices by using matrix factorization. This technique is usually effective because it enables the data scientist to discover the latent features underlying the interactions between customers and products. Therefore, factorization machine models can predict customer ratings on products even when 99% of the cells in the matrix are empty.

Outliers

Outliers can also be detected during this phase, and data scientists must decide whether these data points are erroneous, or if they depict unusual circumstances. Sometimes finding outliers might represent a solution to a business problem. Outliers can describe a problem in the transactional system, which leads the company to improve the data quality efforts, or it can represent an anomaly, a correct value that is very unusual and very unexpected. This outlier can highlight a potential fraud event, a failure in machinery, or a default in financial operations. There are techniques to reduce the effect of the outliers in predictive models and other techniques to eliminate the outliers from the training and validation processes.

Mis-scaled Input Variables

Finally, data scientists should examine the scale of the predictor variables. In some modeling situations, re-scaling the variables to the same scale might be beneficial. For some models, different scales might not represent a big problem, like in regression or tree-based models. But for complex models like neural networks, different scales might represent a bias for the way the model considers and accounts for the distinct inputs. Salary might be more important than age just because its scale is bigger. Therefore, getting all input variables in the same scale (-1 to 1 or 0 to 1) can be very beneficial to neural networks.

An important goal in data exploration is to understand all the possible data sources and predictor variables that are used to solve the business problem based on analytical reports. Unsupervised models like clustering or segmentation, or supervised models like regression and neural networks, or even based on semi-supervised, like text analytics, optimization and network analysis can be used. It is important to understand that the process of exploring the data and analyzing the variables might not just solve business problems along the way, but also it can raise valuable business insights about several topics.

Introduction to Data Preparation

In the data preparation phase, data scientists need to access all the data sources available in the company to solve the business problem. Some data sources reside externally or are provided by third-party companies that can add and enrich the company's information about business problems. The data sources can be customer information, billing information, and transaction information. A common problem is that the data usually consists of multiple rows per customer, like in a data warehouse or data mart. Because of that, all the data must be merged to end up with one single customer version.

Tasks associated with data preparation include the following:

- Representative sampling
- Event-based sampling
- Partitioning the data
- Imputing missing values
- Replacing erroneous variable values
- Transforming variables
- Feature extraction
- Feature selection

Representative Sampling

If the data source comprises many customers, data scientists might want to take a sample of the customers. This might seem of minimal importance in the age of extremely fast computers. However, the model-fitting process occurs only after the completion of a long, tedious, and error-prone data preparation process. Sampling is done for data efficiency purposes because smaller sample sizes for data preparation are usually preferred. Splitting the data into multiple samples enables data scientists to assess how well the models generalize the predictions for future data as we are going to see in data partitioning.

Figure 2.1: Sampling Data

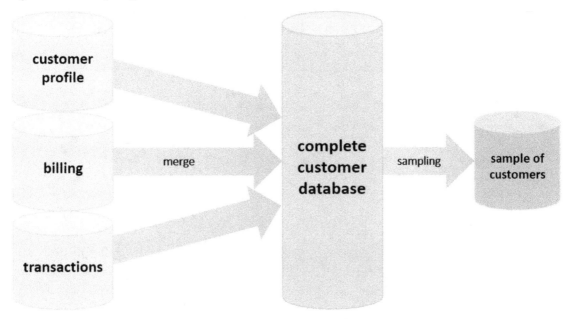

Event-based Sampling

In many business situations, the target that we are trying to model is rare. For example, the churn rate for a telecommunications company might range from 2% to 5%. One common strategy for predicting rare events is to build a model on a sample consisting of all the events and then merge to a sample of the nonevents. The values of some business targets might not relate to the event of interest or to the business goal. For example, the rare event could be a churn event, when the customer decides to quit the company or the product or service. The nonevent is when the customer does not make churn, and when the customer is still in the company, consuming and using some of the products and services. The new sample might have a 50% event rate instead of the original 2% to 5% event rate.

The advantage of event-based sampling is that data scientists can obtain, on average, a model of similar predictive power with a smaller overall case count. This sampling strategy works because the amount of information in a data set with a categorical outcome is determined not by the total number of observations in the data set, but by the number of observations in the rarest outcome category, which is usually the number of events. This sampling approach allows the model to capture both relationships between the inputs and the event of interest and the inputs and the nonevent. If you have in the training data set just nonevents, or at least a vast majority of them, the model tends to easily capture just the relationship of the inputs and the nonevent. Even if the model predicts there is no churn, the model will be correct 99% of the time. This 99% classification rate for this model is a very good classification rate, but the company will miss all the events. The overall model performance will be good, but the company will miss the opportunity to retain possible churners.

Figure 2.2: Event-based Sampling

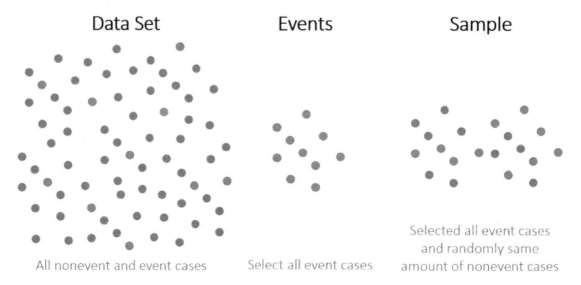

Data Set   Events   Sample

Selected all event cases
and randomly same
amount of nonevent cases

All nonevent and event cases   Select all event cases

Partitioning

Analytical models tend to learn very fast and efficiently capture the relationship between the input variables and target. The problem is they can learn too much. The models can capture almost perfectly the correlation between the inputs and the target, but just for the time frame that it has been trained. As the data changes over time, the model should generalize as much as possible to account for the variability of the data in different time frames. When the model has high predictive accuracy for the trained period but not for future data, it is said that the model overfits the data.

The simplest strategy for correcting overfitting is to isolate a portion of the data for assessment or validation. The model is fit to one part of the data, called the training data set, and the performance is evaluated on another part of the data, called the validation data set. When the validation data are used for comparing, selecting, and modifying models, and the chosen model is assessed on the same data that was used for comparison, then the overfitting principle still applies. In this situation, a test data set should be used for a final assessment.

In situations where there is a time component, the test data set could be gathered from a different time. This would generalize the model even more, as the model should be deployed in production in a different time frame anyway. That means a model is trained based on past events where the target is known. Then once the best model is selected, it will be deployed in production to predict the events in future data, where the target is unknown. The model needs to generalize well to new data to account for variability in the data over time. When a test is made by using data varying in time, it helps the data scientist select the most accurate model but also the one that generalizes best. For example, a model that is fit on data that is gathered from January to June might not generalize well to data that was gathered from July to December. This also raises a problem that there is no model that lasts forever. In some point in time, the model performance will decay especially if the data changes over time. As customers change their behavior, the data used to describe that behavior will change and the model trained and deployed using past data will no longer accurately predict future events.

Figure 2.3: Partitioning

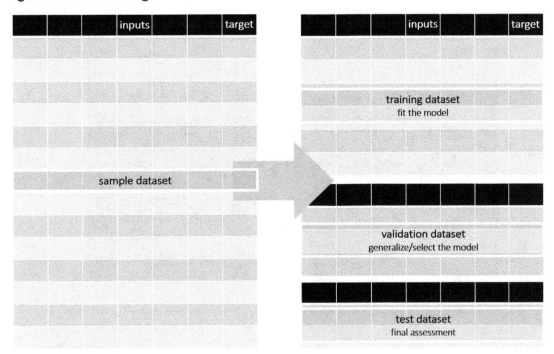

Imputation

The standard approach for handling missing values in most modeling approaches is complete-case analysis, where only those observations without any missing values are used in the training process. That means any observation with a missing value in any of the input variables used to train the model is eliminated from the training process. The problem is even a small number of missing values can cause an enormous loss of data in high dimensions. Imagine an input data set with 300 variables. If one customer has a single missing value in any of these 300 variables, he or she is discarded from the training data set. Reducing the training data because of missing values is one problem. The other problem is to score a model when some values are missing in the future data. Imagine a predictive model in production scoring customers that are likely to make churn or not. One of the input variables is age. In the future data, if the customer being scored has no age value, the model cannot score this customer. For example, in logistic regression, the model cannot multiply the coefficient by a missing value. When data scientists score new customers, customers with missing values will have a missing predictive score such as a probability of having the event. In other words, data scientists cannot generate predictions for customers with missing values in the predictor variables. Therefore, some type of missing value imputation is necessary.

Imputation is a common process to avoid missing values in observations used during training, validation, test, and deployment processes. *Imputation* means filling in the missing values with a reasonable value. Many methods were developed for imputing missing values such as mean, median, mode, and midrange imputation. These values are based on the distribution of the predictor variable.

Figure 2.4: Missing Values and Imputation

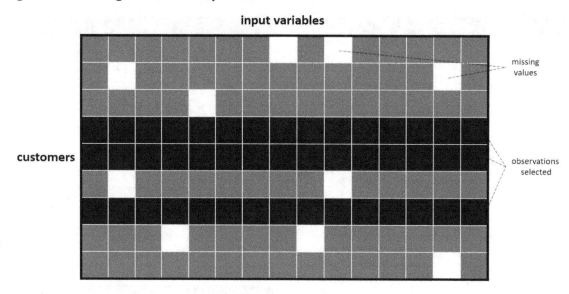

For example, recall the previous example of the customer with the age missing in the training data set. The mean of the age considering all customers in the training data set is used for that customer. Data scientists can also create missing indicator variables that are binary variables that indicate whether the predictor variable value is missing or known for that observation. These variables can be used to capture the relationship between the response variable and the missingness of the predictor. For example, maybe customers with missing incomes have a strong relationship with fraudulent transactions. If a very large percentage of values is missing, then the predictor variable might be better handled by omitting it from the analysis and using the missing indicator variable only.

Figure 2.5: Imputation and Missing Values Indicator

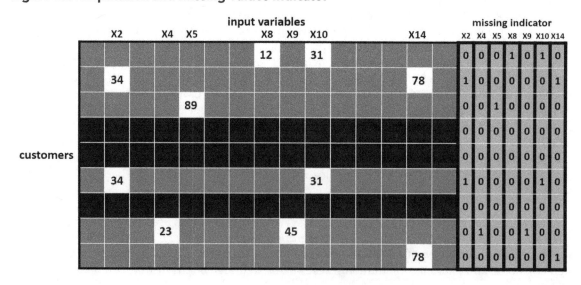

There are other methods to handle missing values like binning the variable into groups and having one group represent all the missing values. The predictor variable would then be treated as a categorical variable in the analysis. Another very efficient method, probably the most accurate approach to handle missing values, is to estimate the missing value based on the other inputs. The data scientist would use another predictive model to estimate the missing values before estimating the target. For example, imagine two customers have the salary missing in the data set. Imputation by mean would add the same salary for both customers. However, let's assume the first customer is 50 years old, director of a company, and owns a home valued at about one million dollars. The second customer is 20 years old, still attending college, and is not a homeowner. These two customers should have very different salaries. However, the imputation by mean would impute the exact same value for both. An imputation model, like a decision tree, would potentially impute very different values for each one of them, as the estimation of the missing salary would be based on the rest of the input variables.

Replacement

In some situations, the data has erroneous values that need to be replaced. For example, suppose the customers in a financial institution data set must be at least 18 years of age. If the data shows customers younger than 18, one remedy is to replace those values with 18. Another remedy is to treat these values as missing and impute the values based on the other predictor variables values in the data, as described in the imputation section. If the percentage of errors in a variable is too high, the data scientists can decide to discard this variable from the analysis.

Transformation

Transformation of the input variables is a common data preparation task. These transformations can be a reasonable method to improve the fit of the model. Data scientists can take a mathematical transformation to create a more symmetric distribution that should reduce the effect of outliers or

Figure 2.6: Mean Imputation Technique

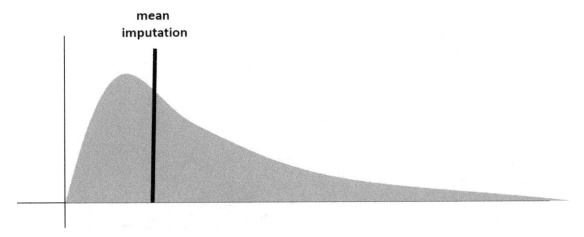

Figure 2.7: Replacement Approach

CustID	Income	Home Value	Age	Gender
1	90,000	450,000	44	M
2	105,000	520,000	48	M
3	78,000	380,000	37	F
4	56,000	260,000	12	M
5	67,000	390,000	-4	F
6	87,000	467,000	39	F
7	98,000	568,000	33	M
8	48,000	278,000	29	F
9	54,000	304,000	0	F
10	120,000	620,000	62	F
11	86,000	513,000	57	M

CustID	...	Age
1		44
2		48
3		37
4		18
5		18
6		39
7		33
8		29
9		18
10		62
11		57

pure replacement

replacement and estimation

CustID	...	Age
1		44
2		48
3		37
4		.
5		.
6		39
7		33
8		29
9		.
10		62
11		57

CustID	...	Age
1		44
2		48
3		37
4		27
5		41
6		39
7		33
8		29
9		32
10		62
11		57

heavy tails in the original variables' distribution. Data scientists can also standardize variables to be in the same scale and range. For example, age and salary can be in the same range such as a scale from 0 to 1, making sure they both equally contribute to the fit of the model. Variable transformations can also help the model capture nonlinear relationships between the inputs and target.

Another transformation is to bin the input variable values and treat it as a categorical variable. This can also reduce the effect of outliers and account for the nonlinear relationship between the target and the input variable. Imagine a data set where most of the customers have a very similar salary, let's say, a low salary, some with medium salary and very few with a high salary. The salary distribution has a huge block of observations at the beginning of the histogram, and then few blocks going down and a long tail for the high salary cases. A distribution like that makes it hard for the model to capture any correlation between the salary and the event of interest such as bad debt.

A log transformation, or a quantile binning, can better distribute the observations into different ranges of salaries and make a possible correlation between the transformed input variable and the target easier to capture as shown in Figure 2.8.

Figure 2.8: Log Transformation Approach

original input scale   log scale

Feature Extraction

Feature extraction, otherwise known as *feature creation*, creates new variables or features from the initial set of variables by combining the original variables. These features encapsulate the important information from multiple original input variables. In other words, feature extraction might reduce the dimensional input space and create a more informative input to the model. This new feature can be more highly correlated to the target than any other original input variable.

In many situations, feature extraction is based on domain knowledge. For example, if the data scientists suspect that the number of phone calls to the call center is related to churn, they might create features such as the number of phone calls in the last six months for a customer or the number of phone calls divided by the total number of phone calls for all customers. These derived variables could be useful predictors in the final model.

Feature creation can be based on text analytics where unstructured text is converted to predictor variables. Principal component analysis (PCA) and autoencoder data transformations can also be used to create new features.

Principal Component Analysis

Principal components are weighted linear combinations of the predictor variables where the weights are chosen to account for the largest amount of variation in the data. Total variation, in this case, is the sum of the sample variances of the predictor variables. The principal components are numbered according to how much variation in the data is accounted for. Since the principal components are orthogonal to each other, each principal component accounts for a unique portion of the variation in the data.

Figure 2.9: Feature Extraction

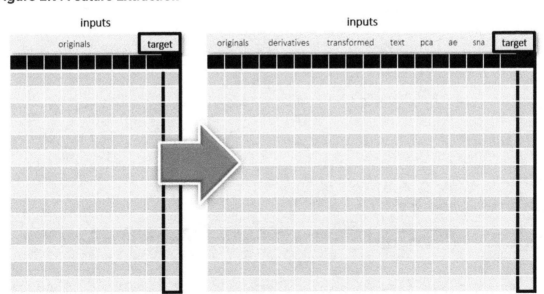

Principal components analysis can be used for dimension reduction since usually only the first few principal components explain enough of the proportion of the total variation in the data. For example, if the data scientists have 1,000 variables but the first 10 principal components explain 95% of the variation in the data, then only the first 10 principal components would be used in the model. It reduces the high dimensionality in the input space, the original variables, and creates more informative variables for the model.

Figure 2.10 graphically illustrates principal component analysis. The first principal component (PC1) is constructed in such a way that it captures as much of the variation in the input variables set as possible. Then the second principal component (PC2) is orthogonal to PC1 and captures as much as possible of the variation in the input data not captured by PC1. The third component follows the same rule and the process keeps going.

Text Mining

Text parsing processes textual data into a term-by-document frequency matrix. Transformations such as singular value decomposition (SVD) alter this matrix into a data set that is suitable for predictive modeling purposes. SVD is simply a matrix decomposition method. When it is used on a document by term matrix, a set of predictor variables is produced that contains information about what is written in the text. The set of coefficients that are produced can be used to derive concepts or topics from the document collection. For example, the documents could be Call Center dialogues and the terms could be angry, frustrated, upset, dislike, dissatisfied, and so on. The coefficients can be used to create variables that represent the strength of the term in the document. Variables that show the strength of terms associated with unhappy customers should be useful predictors in a churn model.

Figure 2.10: Principal Component Analysis

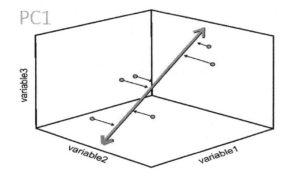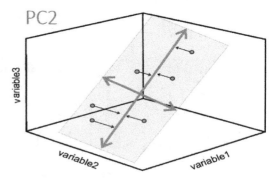

Figure 2.11: Text Mining

	Term 1	Term 2	Term 3	...
Doc 1				
Doc 2				
Doc 3				
...				

		Structured Data				Inputs from Unstructured Data				
ID	Var 1	Var 2	Var 3	...	SVD 1	SVD 2	SVD 3	...	Target
...									

Variable Clustering

Variable clustering finds groups of variables that are as correlated as possible among themselves and as uncorrelated as possible with variables in other clusters. This is used as a dimension reduction technique as the data scientists would choose one variable from each cluster based on subject-matter knowledge. The data scientists could also choose a representative variable from each cluster based on the correlation with its own cluster and the correlation with the other clusters.

Autoencoder

Autoencoder data transformation is an algorithm that aims to transform the predictor variables into derived variables with the least amount of distortion. In other words, it attempts to discover structure within the data to develop a compressed representation of the original variable. The first step is encoding where the algorithm efficiently compresses the data. The second step is decoding where the algorithm tries to reconstruct the original data. The decoder is used to train the encoder.

If the predictor variables were independent of each other, the encoding step would produce derived inputs with very limited value. However, if some sort of structure exists in the data such as correlated predictor variables, then the encoding step would produce derived inputs that reconstruct the predictor variable without holding on to the redundancies within that variable.

Autoencoder data transformation can be used for dimension reduction just like principal components analysis. Whereas principal components analysis attempts to discover linear relationships in the

Figure 2.12: Variable Clustering

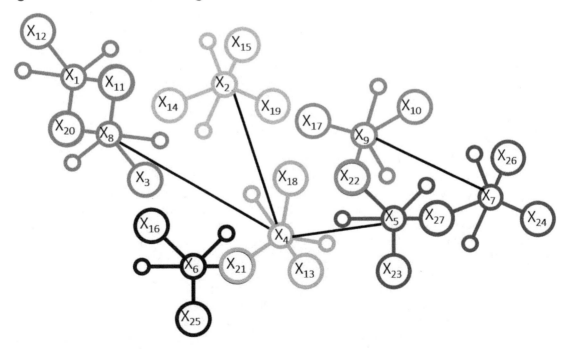

Figure 2.13: Autoencoder Data Transformation

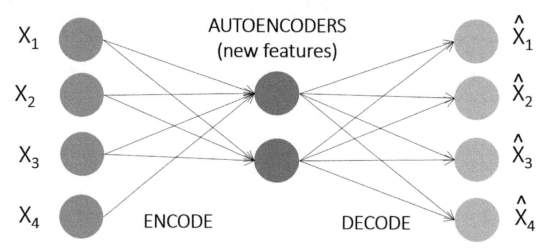

data, the derived variables from autoencoders can discover nonlinear relationships. Furthermore, whereas principal components are uncorrelated, the derived variables from autoencoders might have correlations since autoencoding strives for accurate reconstruction of the variable.

Feature Selection

Data scientists usually deal with hundreds of predictor variables. This limits the ability to explore and model the relationships among the variables since as the dimensions increase, the data becomes sparse. The large number of predictor variables leads to the curse of dimensionality, which means the complexity of the data set increases rapidly with the increased number of variables. The bottom line is the amount of data a data scientist needs increases exponentially as the number of variables increase.

The remedy to the curse of dimensionality is feature selection, also known as dimension reduction. In other words, data scientists want to eliminate irrelevant and redundant variables without inadvertently eliminating important ones. Some of the dimension reduction methods are correlation analysis, regression analysis, and variable clustering.

When features are created, a common recommendation is to eliminate the original variables that were used to construct the features because the features and the original variables will probably be redundant. For example, if the log of age is created, the original age variable is eliminated. However, another point of view is to keep all the variables for the dimension reduction techniques and see which ones survive to the final model. In this way, data scientists see whether age or the log of age survives to the final model.

Redundancy among predictor variables is an unsupervised concept since it does not involve the target variable. On the other hand, the relevancy of a variable considers the relationship between the predictor variable and the target variable. In high-dimensional data sets, identifying irrelevant variables is more difficult than identifying redundant variables. A good strategy is to first reduce redundancy and then tackle irrelevancy in a lower dimension space.

Figure 2.14: Variable Selection

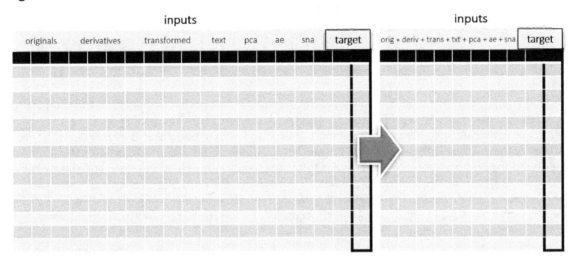

A redundant variable does not give any new information that was not already explained by other variables. For example, knowing the value of input household income usually is a good indication of home value. As one value increases, the other value also increases.

An irrelevant variable does not provide information about the target. For example, if the target is whether you gave to a charitable organization, predictions change with the predictor variable response to previous solicitations, but not with the predictor variable show size.

Figure 2.14 shows a data set with numerous variables such as the original variables, derived variables, transformed variables, variables obtained from text mining, variables obtained from principal component analysis, variables obtained from autoencoder data transformation, and the variables obtained from social network analysis. The goal is to reduce the number of variables down to a reasonable number while still maintaining a high predictive accuracy for the model. It should be noted that social network analysis is discussed in a later chapter.

Model Selection

Once a reasonable number of variables have been selected, data scientists usually have several choices of models to choose based on their complexity. A common pitfall is to overfit the data in which the model is too complex. An overly complex model might be too sensitive to the noise in the data and not generalize well to new data. An example is a model with many unnecessary higher order terms such as age*age and income*income. However, using too simple a model can lead to underfitting where the true features are ignored.

Model Generalization

Typically, model performance follows a straightforward trend. As the complexity of the model increases and more terms are added to the model, or more interactions in the training process, the fit on the training data set generally improves as well. Some of this increase is attributable to the model capturing relevant trends in the data, sometimes very specific correlations between the inputs and the target. However, some of the increase might be due to overfitting the training data as the model might be reacting to random noise. That will work well for the training data set but not for the validation data set or even for the test data set. And probably not for future data. Therefore, the model fit on the validation data for models of varying complexity is also examined.

The typical behavior of the model fit on the validation data is an increase in model fit as the complexity increases, followed by a plateau, followed by a decline in performance. This decline is due to overfitting. Consequently, a common recommendation is to select a model that is associated with the complexity that has the highest validation fit statistic. A recommended practice is to select the simplest model with the best performance. That means select the model with the best performance under the validation data, but the simplest model possible. If the assessment error plot shows a plateau in the error for the validation data set, then select the simplest model at the beginning of the plateau. Simpler models tend to generalize better. Complex models tend to be affected by change in the input data because their structure is more complex.

Figure 2.15: Fit Versus Complexity

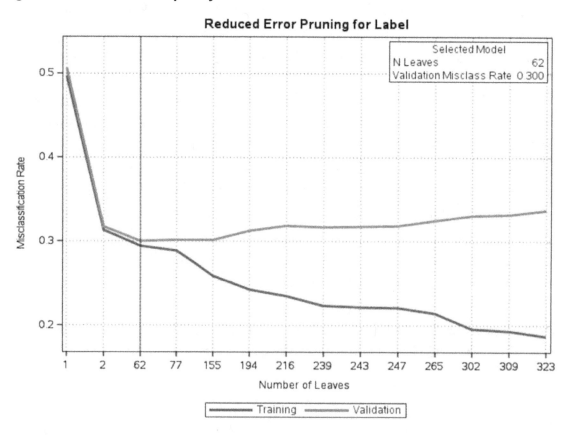

Bias-Variance Tradeoff

The goal of the data scientist is to fit a model with low bias and low variance. *Bias* is the difference between the average prediction of the model and the correct value that we are trying to predict while *variance* is the variability of model prediction for a given data point. A model with high bias misses the important relationships between the predictor variables and the target. This is an example of underfitting. However, a model with high variance models the random noise in the data. This is an example of overfitting. The bias-variance tradeoff is where the data scientists choose a model that accurately depicts the relationships in the data, but also chooses a model that generalizes well to new data.

Evaluating the model on the data the model was fit on usually leads to an optimistically biased assessment. For example, a model might be fit on one set of data, the training data set, and the accuracy of the model was estimated at 70%. However, when the model was applied to a new data set, the validation data set, the accuracy was estimated at 47%. Why the large difference in accuracy percentages? This is an example of overfitting where the model was overly sensitive to peculiarities in the training data set and the model did not generalize well to the validation data set.

Summary

This chapter introduced the main concepts of data exploration, how data exploration is a key stage for any data science project, and how it can address many business issues as a result of reviewing and exploring the information needed to solve a business problem through an analytical model. Data scientists need to understand how important data exploration is in reviewing the information needed to create analytical models to solve business problems. Data exploration tasks help companies fix data quality issues and transform raw data into relevant business information.

This chapter introduced specific concepts about data preparation. Concepts about how to merge data from different sources to create a more comprehensive view about the customers and the business problems were covered. Different methods of sampling, techniques for data partitioning, different approaches to transform the original input data into relevant information to enhance supervised and unsupervised models, and techniques to create new features from the original data to improve model results were covered. The chapter concluded by discussing some statistical measures to select the inputs used in the models and how to select the model that generalizes well to new data.

Chapter 3: Supervised Models – Statistical Approach

Chapter Overview

In this chapter, you will learn about supervised modeling. The term supervised refers to observations or cases where the target is known. Based on the known events, a model is trained to capture the relationship between the inputs and the target. There are several techniques and algorithms used in supervised modeling, including statistical models and machine learning models. Chapter 3 covers the statistical models, including linear and logistic regression models and decision trees. Chapter 4 covers the machine learning models.

The main goals of this chapter include:

- Explain the purpose of using statistical supervised models, focusing on linear regression, logistic regression, and decision trees.
- Describe several types of supervised models, the distinct techniques, or algorithms to be used, and the circumstances under which one of them might be used to solve a business problem.
- Interpret the results of the supervised models. Linear regression, logistic regression, and decision tree models are all interpretable. This interpretability can be used to drive business actions based on the models results.

Classification and Estimation

Classification and estimation are common types of predictive models. Classification assumes that the target is a class variable. The target can be a binary class, 0 or 1, yes or no, or it can be a multinomial class, like 1, 2, 3, 4, and 5, or low, medium, and high. For example, is this business event a fraudulent transaction (yes or no)? Estimation assumes that the target is a continuous number. The target can take any value in the range of negative infinity to positive infinity.

Both predictive models, classification and estimation, require the following:

- Observations/cases/instances:
  - A real case comprising a set of attributes that describe the observation.

- Inputs/attributes/variables:
  - o The measures of the observation, or its attributes. It can be demographic information about the customer, such as age, salary, and marital status.
- Target/class/label:
  - o A tag or label for each observation. Default or no default is an example.

The data used to train a supervised machine learning model consists of a set of cases or observations that happened in the past, which means the value of the target is known. The main idea is to find relationships between the predictor variables and the target. For example, looking at all churn cases that happened in the past six months, what input variables can explain whether the event of churn and non-churn occurred? If a model can be trained based on past cases and find what input variables can explain the churn and the non-churn, then this model can be used to predict cases of churn in the future.

A statistical model maps the set of input variables to the target. The model tries to create a concise representation of the inputs and the target. It tries to capture the relationship between the inputs and the target. For example, when the target is a binary target such as churn versus no churn, the model tries to explain, based on the input variables, whether the customer is willing to make churn or not. What input variables are associated to the yes (churn) and what values? What variables are associated to the no (no churn) and what values? The target is the outcome to be predicted. The cases are the units on which the prediction is made.

For example, imagine a company selling various products to its customers over time. A sample of customers can be identified with some important characteristics, including:

- Age
- Gender
- Average revenue generated over the years
- How long the customer has been in the database
- Number of products purchased
- Marital status
- If the customer is a homeowner
- How many children the customer has

A data scientist can analyze this sample containing past transactions and search for some correlations between the customer's characteristics and the event of purchasing a new product. For example, what are the customer's characteristics when they purchased a new product in the past? What are these characteristics when the customer did not purchase the new product in the past? Imagine also that the data scientist can find a clear correlation between these explanatory variables and the event of purchasing a new product or not. Let us say, looking at the past events, the customers who purchased the new product present some unique characteristics, such as they are on average 46 years to 48 years old. It does not matter what the gender is as both male and female equally purchased the new product. They spend on average USD 270 per month, varying from USD 234 to USD 310. They are customers for seven years on average, varying from six to eight. They have two products on average, varying from two to three products. They are married or divorced. They are all homeowners. They all have kids, on average two children, varying from two to three. A data scientist can think of these relationships as a mathematical equation to classify new cases.

For example, if a new event takes place, let us say, a sales campaign for old customers who have not purchased this new product. One customer is 49 years old, generating on average a monthly revenue of USD 297, has been a customer for 5 years, has already purchased one product, is married, is a homeowner, and has three children. Following that mathematical equation, a data scientist can easily compare the characteristics of this customer with all the customers who purchased the new product in the past and infer that this customer has a high likelihood to purchase the new product. The sales campaign can then target this customer.

Another customer presents a different set of characteristics, such as 39 years old, USD 110 average revenue, eight years as a customer, purchased one product, is single, is a homeowner, and has no kids. Based on that mathematical equation, a data scientist can infer that this customer has a low probability to purchase the new product and then does not target them in the sales campaign.

By doing this, the company will target just the customers with a high propensity to purchase the new product. This will save a substantial amount of money for the company in operationalizing the campaign, and it will dramatically increase the response rate of the business event. Of course, some customers targeted in the campaign will eventually not purchase the new product, and some customers not targeted in the campaign would purchase the new product. This is the intrinsic error associated with the model. Customers with similar characteristics behave very differently. Errors are associated with all models.

The term supervised is used when the target variable value is known for each case. The known target variable value is used to train the model in finding the relationships between the inputs and the target. Remember the case that we were looking at before. We were trying to understand the relationship between the customer's characteristics and the event of purchase. How we could use all the customer's characteristics to explain the past events of purchase and then use these relationships to predict future events, or what customers would be more inclined to purchase a new product?

Figure 3.1: Supervised Model Training Method

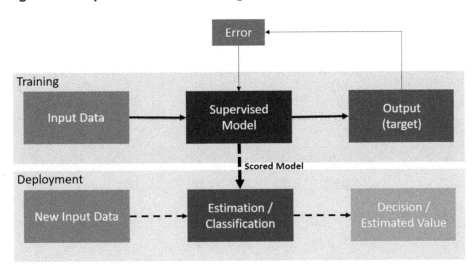

All the models we are going to see in Chapters 3, 4, and 5 use the exact same approach. Linear regression, logistic regression, decision tree, neural network, forest, gradient boosting, and support vector machines all use past data to create a concise map to estimate the relationships between the input variables to the target values. In other words, the models estimate the relationships between the customer's characteristics to the event of purchase and no purchase. Once the relationships are identified, and the map is built, the model can be used to predict future events, or to select which customers who do not have the product would be more inclined to purchase it. Based on that map, which can be a formula, an equation, or a set of rules based on thresholds, the company can identify all customers who are more likely to purchase the product and approach them in a specific sales campaign. All supervised models behave very similarly, using the past known events to estimate the relationships between the predictor variables and the target. If a good model is found, that model can be used to predict future events.

Because the purpose of the supervised model is to predict the unknown values of the target, the main goal of the supervised model is to generalize to new data or future cases. Generalization means the ability to predict the outcome on new cases when the target is unknown. That is the reason we use different data sets to train the model (capture the relationship between the inputs and the target) and validate the model (generalize the model to make good predictions on the new data).

Generalization is also involved in model assessment. The model is fit to the training data set, and the performance is evaluated on the validation data set by comparing the predicted values to the observed values of the target. Since the target value is known, the assessment is straightforward, comparing the values the model predicts for the target and the values observed for the target.

In this chapter, we are going to discuss three distinct types of models:

- Linear Regression – used to predict continuous targets. One assumption of the linear regression model is that there is a linear relationship between the inputs and the target.
- Logistic Regression – used to predict discrete targets such as binary, ordinal, and nominal outcomes. One assumption of logistic regression is that there is a linear relationship between the inputs and the logits.
- Decision Trees – used to predict both continuous and categorical targets. There is no assumption for the relationships between the inputs and the target. Decision trees are universal approximators because theoretically they can capture any type of relationship between inputs and target.

Linear Regression

The relationship between the target and the input variables can be characterized by the equation:

$$y_i = \beta_0 + \beta_1 x_i + \varepsilon_i, i = 1...,n$$

where

y_i is the target variable.

x_i is the input variable.

β_0 is the intercept parameter, which corresponds to the value of the target variable when the predictor is 0.

β_1 is the estimate (slope) parameter, which corresponds to the magnitude of change in the target variable given a one-unit change in the input variable.

ε_i is the error term representing deviations of y_i about $\beta_0 + \beta_1 x_i$.

Estimates of the unknown parameters β_0 and β_1 are obtained interactively by the method of ordinary least squares. This method provides the estimates by determining the line that minimizes the sum of the squared vertical distances between the observations and the fitted line. In other words, the fitted or regression line is as close as possible to all the data points.

In a linear regression, the estimation for the target variable is formed from a simple linear combination of the inputs. The intercept is the average value of the target when the predictor values are zero, and the remaining parameter estimates determine the slope between each input and the target. The prediction estimates can be viewed as a linear approximation to the expected value of a target conditioned on observed input values.

One of the greatest strengths of the linear regression model is its simplicity, for both implementing and interpreting. Linear regression is quite simple to implement since it is only a linear equation. The coefficients from the model are easily interpretable as the effect on the target given a one-unit increase in the predictor variable controlling for the other predictor variables. In most cases, the results of the model can be obtained quickly, and new observations can be scored quickly. In other words, predictions on new cases can be easily obtained by plugging in the new values of the predictor variables.

The drawbacks of the linear regression model are the models are limited to *normally distributed residuals*, which are the observed values minus the predicted values. If the residuals show a non-normal distribution such as a skewed distribution, then a generalized linear regression model might be helpful.

Another drawback is that a high degree of collinearity among the predictors can cause model instability. The instability of the coefficients might not lead to a simple model that identifies the predictor variables that are the best predictors of the target. Therefore, data scientists should reduce redundancy first if they want to use linear regression models.

If there are non-linear relationships between the predictor variables and the target, data scientists will have to add higher order terms to properly model these relationships or perform variable transformations. Finding the appropriate higher order terms or variable transformations can be time-consuming and cumbersome.

Finally, linear regression models can be affected by outliers that can change the model results and lead to poor predictive performance.

Use Case: Customer Value

An example of a business problem that can be addressed using linear regression is predicting the amount of earned revenue by customer. The results of the model can drive sales campaigns to push customers with low earned revenue but a high predicted level of spending. On the other hand, the model can indicate customers who are close to the possible maximal earned revenue and then they can be removed from non-critical sales campaigns, avoiding eventual attrition to them. There is no reason to add to some sales campaign customers who have reached the possible maximal earned revenue. By removing them from the non-critical campaigns, the company can reduce the operational cost to contact those customers and can also avoid inundating good customers with unnecessary marketing advertisements.

The input variables or predictors in the data are:

- Demographic information about the customers
- Product usage information such as frequency, recency, number of products, value per product, total value, revenue by product, total revenue, amount of usage per product, and total usage
- Payment type, day, frequency of payment, amount paid, and so on
- Payment delay information
- Other products, bundles and services consumed (not in the campaign)
- Aging of the customer in the company/product/service
- Others

The target variable is:

- Maximal earned revenue by customer

Logistic Regression

Logistic regressions are closely related to linear regressions. In logistic regression, the expected value of the target is transformed by a link function to restrict its value to the unit interval. In this way, model predictions can be viewed as primary outcome probabilities between 0 and 1. A linear combination of the inputs generates a *logit score*, or the log of the odds of the primary outcome, in contrast to linear regression, which estimates the value of the target. The range of logit scores is from negative infinity to positive infinity. For binary prediction, any monotonic function that maps the unit interval to the real number line can be considered as a link. The logit link function is one of the most common. Its popularity is due, in part, to the interpretability of the model.

For example, if you want to use logistic regression for classification of a binary target, you would want to restrict the range of the output to be between 0 and 1. The logit link function transforms the continuous logit scores into probabilities between 0 and 1. The continuous logit scores, or the logit of $\hat{p}$, is given by the *log of the odds*, which is the log of the probability of the event divided by the probability of the non-event. This logit transformation transforms the probability scale to the real line of negative infinity to positive infinity. Therefore, the logit can be modeled with a linear combination since linear combinations can take on any value.

The logistic model is particularly easy to interpret because each predictor variable affects the logit linearly. The coefficients are the slopes. Exponentiating each parameter estimate gives the odds ratios, which compares the odds of the event in one group to the odds of the event in another group.

The odds ratio shows the strength of the association between the predictor variable and the target variable. If the odds ratio is 1, then there is no association between the predictor variable and the target. If the odds ratio is greater than 1, then the group in the numerator has higher odds of having the event. If the odds ratio is between 0 and 1, then the group in the denominator has higher odds of having the event. For example, an odds ratio of 3 indicates that the odds of getting the event for the group in the numerator are three times that for the group in the denominator. The group in the numerator and denominator is based on the coding of the input variable. For example, if the parameter estimate for the input variable age is 0.97, then the exponent is 2.66. That means, for a one unit increase in age, the odds are increased by 166% ((2.66 - 1)*100).

In general, the predictions from logistic regression can be rankings, decisions, or estimates. Analysts can rank the posterior probabilities to assess observations to decide what actions to take. For example, in a collection campaign, the customers can be contacted based on the rank of the posterior probability. Higher posterior probabilities mean that customers are more likely to pay their bills. To get a decision, you need a threshold. The easiest way to get a meaningful threshold is to convert the prediction ranking to a prediction estimate. You can obtain a prediction estimate using a straightforward transformation of the logit score, the logistic function. The logistic function is the inverse of the logit function. You can obtain the logistic function by solving the logit equation for p. The threshold can be the value for a random event, for example, 0.5 or 50%. For example, the odds to flip a coin and get a tail or a head is 50-50. If the posterior probability is greater than 0.5, then the decision is that the observation likely has the event of interest. For example, all customers with the posterior probability greater than 0.5 are likely to pay their bills. All customers with a posterior probability less than or equal 0.5 are likely to not pay their bills.

The parameter estimates in a logistic regression are commonly obtained by the method of maximum likelihood estimation. These estimates can be used in the logit and logistic equations to obtain predictions.

If the target variable is dichotomous (two categories), the appropriate logistic regression model is binary logistic regression. If there are more than two categories (levels) within the target variable, then there are two possible logistic regression models:

- If the target variable is nominal, the appropriate model is nominal logistic regression.
- If the target variable is ordinal (rankings), the appropriate model is ordinal logistic regression.

The binary logistic regression model assumes that the logit of the posterior probability is a linear combination of the predictor variables. The parameters β_0, β_1,...,β_k are unknown constants that must be estimated from the data (k is the number of predictor variables).

The strengths of the logistic regression model are similar to the linear model. First, the model is quite simple to implement since it is only a linear equation. Second, the coefficients from the model are easily interpretable as the effect on the logit given a one-unit increase in the predictor variable controlling for the other predictor variables. When the parameter estimates are exponentiated, odds ratios are obtained that show the strength of the relationship between the predictor variables and the target. Furthermore, the predicted values have a nice probabilistic interpretation. Third, in most cases, the results of the model can be obtained quickly. Fourth, new observations can be scored quickly.

The drawbacks of logistic regression are like linear regression in that a high degree of collinearity among the predictors will cause model instability. Furthermore, if there are nonlinear relationships between the predictor variables and the target, data scientists will have to add higher order terms or variable transformations to properly model these relationships. Finally, logistic regression models can be affected by outliers that can change the model results and lead to poor predictive performance.

Use Case: Collecting Predictive Model

An example of a business problem that can be addressed using logistic regression is the probability of debt repayment. The main goal of the model is to rank insolvent customers based on the probability of paying off their unpaid bills. The results of the model can be used to target the insolvent customers who are more likely to pay off their debt. The input variables in the data are:

- Demographic information about the customers
- Payment type, day, frequency of payment, amount paid, and so on
- Payment delay information
- Aging of the customer in the company, product, or service
- Credit history for the customer
- Past delinquent bills for the customer
- Debt to income ratio, debt to bill ratio, total debt to total bill ratio, and so on
- Others

The target variable is:

- Whether the customer has paid off the unpaid bill

Telecommunications companies usually rank insolvent customers by how much they owe or by the age of the unpaid bills. Then, they target the customers with the largest or oldest debt. However, if the telecommunications company ranked the customers based on the probability of payment and targeted the customers with the highest probabilities, there might be an increase in revenue. This problem can cause even more damage in countries with unstable economic situations and high inflation rates. If customers do not pay their bills, the company needs to get money from the financial market to maintain the cash flow. This money costs much more than the company charges the insolvent customers in terms of fees and interest rates. The longer the customers remain insolvent, more money needs to be collected from the financial market. The main goal for this model is to allow companies to anticipate cash by contacting customers who are likely to pay their unpaid bills first.

Decision Tree

Decision trees are statistical models designed for supervised prediction problems. Supervised prediction encompasses predictive modeling, pattern recognition, discriminant analysis, multivariate function estimation, and supervised machine learning. A decision tree includes the following components:

- An internal node is a test on an attribute.
- A branch represents an outcome of the test, such as color=purple.
- A leaf node represents a class label or class label distribution.
- At each node, one attribute is chosen to split the training data into distinct classes as much as possible.
- A new instance is classified by following a matching path to a leaf node.

The model is called a decision tree because the model can be represented in a tree-like structure. A decision tree is read from the top down starting at the root node. Each internal node represents a split based on the values of one of the inputs. The inputs can appear in any number of splits throughout the tree. Cases move down the branch that contains its input value. In a binary tree with interval inputs, each internal node is a simple inequality. A case moves left if the inequality is true and right otherwise. The terminal nodes of the tree are called leaves. The leaves represent the predicted target. All cases reaching a leaf are given the same predicted value. The leaves give the predicted class as well as the probability of class membership.

Decision trees can also have multi-way splits where the values of the inputs are partitioned into disjoint ranges.

When the target is categorical, the model is called a classification tree. A classification tree can be thought of as defining several multivariate step functions. Each function corresponds to the posterior probability of a target class. When the target is continuous, the model is a called a regression tree. The leaves give the predicted value of the target. All cases that reach a leaf are assigned the same predicted value. Cases are scored using prediction rules. These prediction rules define the regions of the input space in which the predictions are made. Each prediction rule tries to make the region of the input space purer with regard to the target response value.

To illustrate decision trees using business data, a generic data set containing information about payment is used with a binary target of default. For simplicity, the input variables are:

- Previous delay: the number of previous delays since the time analyzed.
- Over billing: the billing amount difference, or the billing amount divided by average billing amount.
- Aging: the time since the customer first started consuming products or services from the company.

The target variable is:

- Bad debt: whether the customer defaulted on their bills or not.

A decision tree is read from the top down starting at the root node. In this example, previous delay is the root node where cases with a value of 0 go to the left in the tree and cases with values of 1 or more go to the right in the tree. (See Figure 3.2.) Previous delay was chosen for the root node based on a split-search algorithm that finds the predictor variable that gives the most significant split between the values of the target variable.

Each internal node represents a split based on the values of one of the predictor variables. The predictor variables can appear in any number of splits throughout the tree. Cases move down the branch that contains its predictor value.

The leaves represent the predicted target. All cases reaching a leaf are given the same predicted value. This value is based on the target value of cases in the training data that reach this leaf. In this example, customers with previous delays in payments, with a billing difference of 10 percent and greater and who have been customers less than 5 years, are predicted as a default.

The strengths of decision trees are that they are easy to implement, and they are very intuitive. In fact, the results of the model are quite easy to explain to non-technical personnel. Decision trees can be fit very quickly and can score new customers very easily.

Unlike linear models such as linear regression and logistic regression, decision trees can handle non-linear relationships between the target and the predictor variables without specifying the relationship in the model.

Missing values are handled because they are part of the prediction rules. For example, if income is missing, the customers can be put into their own input space rather than eliminated from the analysis.

Figure 3.2: Decision Tree Modeling Approach

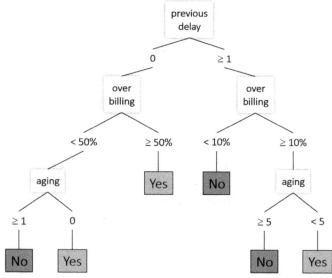

Decision trees are also robust to outliers in the predictor variable values and can discover interactions. An interaction occurs when the relationship between the target and the predictor variable changes by the level of another predictor variable. For example, if the relationship between the target and income is different for males compared to females, decision trees would be able to discover it.

Decision trees confront the curse of dimensionality by ignoring irrelevant predictor variables. However, decision trees have no built-in method for ignoring redundant predictors. Since decision trees can be fitted quickly and have a simple structure, this is usually not an issue for model creation. It can be an issue for model deployment though, in that decision trees might arbitrarily select from a set of correlated predictor variables. To avoid this problem, it is recommended that the data scientists reduce redundancy before fitting the decision tree.

As a drawback, decision trees are very unstable models, which means that any minor changes in the training data set can cause substantial changes in the structure of the tree. The overall performance, or accuracy, can remain the same, but the structure can be quite different. If the structure is different, the set of rules based on thresholds are also different. The common method to mitigate the problem of instability is to create an ensemble of trees. We will see this approach in the next chapter.

Use Case: Subscription Fraud

A business problem where decision tree models can be useful is subscription fraud. In telecommunications, subscription fraud is when a fraudster uses a stolen or a synthetic identity to acquire mobile devices and services with no intention to pay. In many countries, telecommunications regulations allow customers to remain insolvent for a period without getting their services blocked. This causes major financial damages to the companies. Subscription fraud in telecommunications can be even worse as the proceeds and services are sometimes used by organized crime and terrorist networks. The main goal of the model is to detect subscription fraud and to prevent intentional bad debts. Fraud analysts need to be careful when assessing the cases to avoid adversely impacting the customer journey for the genuine customers. Blocking genuine communication services by mistake is a genuine problem.

As shown in Figure 3.3, a usual framework involving fraud – either subscription fraud or usage fraud – consists of a customer relationship management (CRM) system to receive customers' orders. These orders are evaluated by a credit system (it can be accomplished using a credit bureau). In parallel, these orders can also be analyzed by a subscription fraud system, which normally receives information about past customers' transactions. For example, in telecommunications, all raw transactions (calls or even calls attempted) are fetched by the collection systems. This system sends all transactions to a mediation system to aggregate all information and filter the billable transactions. These billable transactions are sent to the billing systems, which process the bills and charge the customers. All this information, in different levels, are used to evaluate and detect subscription and usage fraud. Historical customer information and transaction information are gathered in the data warehouse, which provides the data needed by the data mining tool, environment, or system to train, evaluate, and deploy the predictive models.

Figure 3.3: Systems Framework Involved in Fraudulent Events

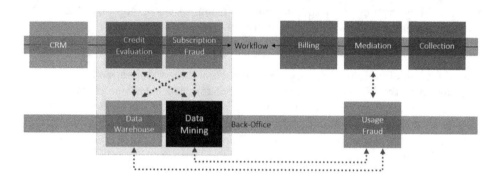

When a service order is placed in the call center (CRM), the service representative must decide in a matter of seconds whether the request is a fraudulent event or a genuine request. This can be accomplished using decision tree models that recognize patterns associated with subscription fraud. Some of the information used could include stolen identities, fake addresses, specific payment methods, and known blocked lists. The models are used to compute the probability of subscription fraud, and these scores are relayed to the service representative. With this information, the service representative can decide whether this is subscription fraud or a genuine customer. As the decision tree models are fit because of a set of rules based on thresholds, some of the rules that are the most correlated to the subscription fraud can be communicated to the representative when evaluating the case. The list of rules generated by the decision tree is especially useful to the service representative during the customer call but also to the team of fraud analysts when analyzing the cases afterward. Some of the high probability subscription fraud cases might go through during the customer service call, but fraud analysts can evaluate cases afterward to decide what actions to take on some of the orders.

Figure 3.4: Information Flow in Fraudulent Events

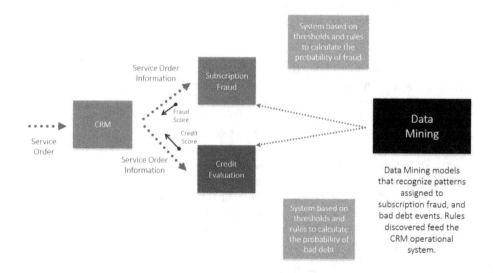

Summary

This chapter introduced the fundamental concepts of supervised modeling. Supervised learning models consider observations where the target is known. Based on the known events, a model is trained to capture the relationship between the inputs and the known target. The statistical models described in this chapter were linear and logistic regression models and decision trees.

Chapter 4: Supervised Models – Machine Learning Approach

Chapter Overview

In this chapter you will learn about supervised machine learning models. Chapter 3 covered supervised statistical models such as linear regression, logistic regression, and decision trees. Chapter 4 covers machine learning models including random forests, gradient boosting models, and artificial neural networks.

The main goals of this chapter are:

- Identify situations that could benefit from using machine learning supervised models.
- Identify some advantages in terms of performance, flexibility, and accuracy, and some disadvantages like the lack of interpretability.
- Describe diverse types of machine learning supervised models and the circumstances under which they might be used, particularly focusing on random forests, gradient boosting, and artificial neural networks.
- Interpret some of the outcomes from supervised machine learning models.

Supervised Machine Learning Models

We are going to discuss three diverse types of models:

- Random Forest – used to predict nominal targets or estimate continuous values. There are no assumptions for the relationships between the inputs and the target. The random forest model is based on multiple independent decision tree models.
- Gradient Boosting – used to predict nominal targets or estimate continuous values. There are no assumptions for the relationships between the inputs and the target. Gradient boosting is based on a sequence of decision tree models.
- Neural Networks – used to predict nominal targets or estimate continuous values. This model is based on a linear combination of nonlinear multiple regressions. The combination of multiple inputs and hidden neurons allows the model to account for nonlinearities between the inputs and the target.

All three models are universal approximators because theoretically they can capture any type of relationship between the inputs and the target.

Ensemble of Trees

Decision trees are particularly good models, easy to implement, fast to train, and easy to interpret, but they are unstable. In some cases, a change in the class label of one case could result in a completely different structure for the tree, even though it has nearly the same accuracy. This can be problematic when deploying decision trees to support business actions. Even though the overall accuracy can remain similar when the final structure of the decision tree changes, the set of rules based on thresholds will be different because they are based on the tree structure. This might impact the business campaign because it is based on that set of rules.

The decision tree's instability results from the considerable number of univariate splits and fragmentation of the data. At each split, there are typically many splits on the same predictor variable or different predictor variables that give similar performance. For example, suppose age is split at 45 since it is the most significant split with the predictor variable and the target. However, other splits at 38 or 47 might be almost as significant. A slight change in the data can easily result in an effect that can cascade and create a different tree. A change in the input data can result in a split at 38 years old. Even more problematic, the change in the input data can result in another input variable being more significant to the target such as income. Then, instead of splitting the input data based on age, the decision tree starts splitting the input space based on income. The final tree structure will be quite different, and therefore, the set of rules and thresholds will also be different.

Several methods have been used to take advantage of this decision tree instability to create models that are more powerful, especially in terms of generalizing the final predictive model. One of the most popular methods is to create an ensemble of decision trees.

An ensemble model is the combination of multiple models. The combinations can be formed in these ways:

- voting on the classifications
- using weighted voting, where some models have more weight
- averaging (weighted or unweighted) the predicted values

There are two types of ensemble of trees: random forests and gradient boosting models.

Random Forest

In a random forest model, the training data for each decision tree is sampled with replacement from all observations that were originally in the training data set. *Sampling with replacement* means that the observation that was sampled is returned to the training data set before the next observation is sampled, and therefore, each observation has the chance of being selected for the sample again. Furthermore, the predictor variables considered for splitting any given decision tree are randomly selected from all available predictor variables. Different algorithms randomly select a specific number of predictor variables. For example, at each split point, only a subset of predictors equal to the square root of the total number of predictors might be available. Therefore, each decision tree is created on a sample of predictor variables and from a sample of the observations. Repeating this process many times leads to greater diversity in the trees. The final model is a combination of the individual decision trees where the predicted values are averaged. Forest models usually have improved predictive accuracy over single decision trees because of variance reduction. If individual decision trees have low bias but high variance, then averaging them decreases the variance.

Random forest is based on a concept called bagging. Bagging is a bootstrap aggregation. It is the original perturb and combine method developed by Breiman in 1996. The main idea of the perturb and combine method is to take the disadvantage of decision trees in terms of instability and turn it into a major advantage in terms of robustness and generalization. These are the main steps:

- Draw K bootstrap samples.
  - A bootstrap sample is a random sample of size n drawn from the empirical distribution of a sample of size n. That is, the training data are resampled with replacement. Some of the cases are left out of the sample, and some cases are represented more than once.

- Build a tree on each bootstrap sample.
  - Large trees with low bias and high variance are ideal.

Figure 4.1: Tree-based Models: Forest

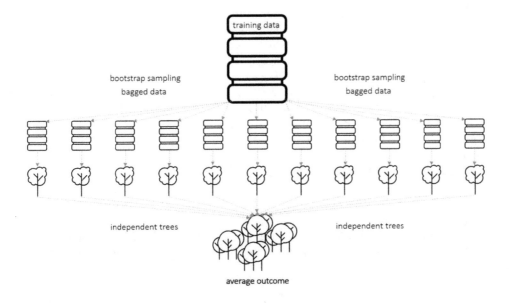

- Vote or average.
  - o For classification problems, take the mean of the posterior probabilities or take the plurality vote of the predicted class. Averaging the posterior probabilities gives a slightly better performance than voting. Take a mean of the predicted values for regression.

Gradient Boosting

Another approach to ensemble of decision trees is gradient boosting. Gradient boosting is a weighted linear combination of individual decision trees. The algorithm starts with an initial decision tree and generates the residuals. In the next step, the target is the residuals from the previous decision tree. At each step, the accuracy of the tree is computed, and successive trees are adjusted to accommodate previous inaccuracies. Therefore, the gradient boosting algorithm fits a sequence of trees based on the residuals from the previous trees. The final model also has the predicted values averaged over the decision trees.

Just like the forest models, the gradient boosting model should have improved predictive accuracy because of variance reduction. It is hoped that the final model has low bias and low variance.

A major difference between random forest and gradient boosting is in the way the ensemble of decision trees is created. In forests, each decision tree is created independently. In gradient boosting, the set of decision trees is created in a sequence. This difference can allow random forest models to be trained faster and gradient boosting models to be more accurate. On the other hand, random forests can better generalize, and gradient boosting models are easier to overfit.

Gradient boosting is based on a slightly different approach than random forests, particularly assigned to the perturb and combine method. Boosting is a machine learning ensemble meta-algorithm for primarily reducing variance. The term boosting refers to a family of algorithms that can convert weak learners (in this case, decision trees with large residuals) into strong learners.

Adaptive resampling and combining methods are examples of boosting. They sequentially perturb the training data based on the results of the previous models. Cases that are incorrectly classified are given more weight in subsequent models. For continuous targets, from the second decision tree in the sequence onwards, the target is the residual of the previous tree. The objective is to minimize the objective function which in this model minimizes the residuals.

The gradient boosting model is shown below.

$$F_M(x) = F_0 + \beta_1 T_1(x) + \beta_2 T_1(x) + \cdots + \beta_M T_M(x),$$

where M is the number of iterations, or decision trees in the gradient boosting model, F_0 is the initial guess, β is the weight for each decision tree in the linear combination, and T is the decision tree model in each iteration.

The strengths of the forest and gradient boosting models are that the models have most of the advantages of decision trees without the model instability and overfitting. The ensemble of tree models

Figure 4.2: Tree-based Models: Gradient Boosting

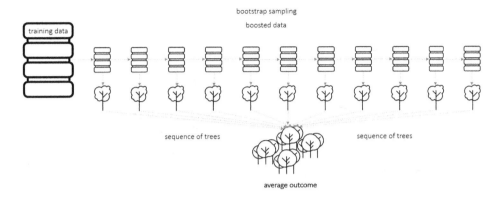

have low bias and low variance, can handle non-linear relationships with the target, can handle missing values, and are robust to outliers.

The weaknesses of the ensemble of tree models are that the simple interpretation of a single decision tree is lost. Forest and gradient boosting models are not simple and interpretable. They also require far more computer resources compared to the single decision tree. Gradient boosting models can train slower than forests as the trees are built in series, where one tree depends on the results of the prior trees.

Use Case: Usage Fraud

A business problem where forest or gradient boosting models can be extremely helpful is usage fraud. Telecommunications usage fraud is when a fraudster uses products and services provided by the telecommunications company with no intention to pay, or even worse, when fraudsters sell or resell telecommunications products and services in the black market. This is a critical concern as sometimes this type of fraud is associated with organized crime and terrorist networks. Although the main goal of the model is to detect usage fraud and to prevent intentional bad debts, the fraud detection team needs to be careful that it does not adversely impact the customer journey for good customers.

Usage fraud might have specific characteristics in terms of calling behavior such as calling specific cities and countries, the duration of the calls, the frequency of the calls, and the recency of the calls. A transactional system processing all calls, called mediation systems, can feed another transactional system to handle fraud detection such as a fraud management system (FMS). This application usually requires a model to learn and identify suspicious behavior and trigger alerts so that the fraud analysts can further investigate the cases. This usage behavior is very often a cumulative behavior over time.

Different machine learning models such as forest and gradient boosting can address this business issue properly. These models are used to compute the probability of usage fraud. If the probability is above a threshold, an alert can be raised. The fraud management system continuously uses these models to score the data to generate alerts. The data scientist can also select only the outcomes

Figure 4.3: Information Flow in Fraudulent Events

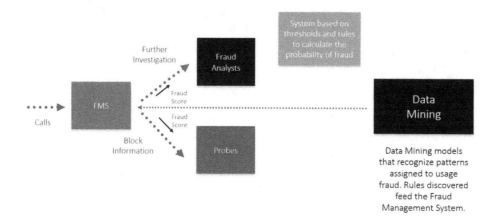

with high probabilities to be transferred to the FMS. This approach aims to avoid false-positive identification alerts to the fraud analysts that will unnecessarily block products and services associated to good customers.

In Figure 4.3, real-time transactions (for example, calls and text messages in telecommunications) feed the FMS. These transactions are used in rules that accumulate customers' behaviors in a recency and frequency perspective. Historical transactions are used in data mining initiatives (classification and estimation supervised models, and clustering and segmentation unsupervised models) to identify all possible rules to detect suspicious transactions. These rules are evaluated and deployed into the FMS. Once an alert is raised, there are basically two options, mostly depending on the type of the rule and its accuracy. If a high accurate alert is raised, an automated system can automatically drop the call by using probes in the network system. If a medium accurate alert is raised, the case can be sent to a fraud analyst for further investigation.

Neural Network

If the relationship between the input variables and the target is nonlinear but it is possible to specify a hypothetical relationship between them, a parametric nonlinear regression model can be built. When it is not practical to specify the hypothetical relationship, a nonparametric regression model is required.

Nonlinear regression models are more difficult to estimate than linear models. Data scientists must specify the full nonlinear regression expression to be modeled and an optimization method to efficiently search for the parameters. Initial parameter estimates also need to be provided and are critical to the optimization process. Another option is a nonparametric regression model that has no functional form and, therefore, no parameters.

Traditional nonlinear modeling techniques are more difficult to define as the number of inputs increase. It is uncommon to see parametric nonlinear regression models with more than a few inputs, because deriving a suitable functional form becomes increasingly difficult as the number of inputs increases. Higher-dimensional input spaces are also a challenge for nonparametric regression models.

Neural networks were developed to overcome these challenges. Although neural networks are parametric nonlinear models, they are like nonparametric models in one way: neural networks do not require the functional form to be specified. This enables data scientists to construct models when the relationships between the inputs and the target are unknown.

However, like other parametric nonlinear models, neural networks do require the use of an optimization process with initial parameter estimates. Furthermore, unlike nonlinear parametric regression models and nonparametric regression models, neural networks perform well in sparse, high-dimensional spaces.

Both regressions and neural networks have similar components but with different names. Instead of an intercept estimate, a neural network has a bias estimate. Instead of parameter estimates, a neural network has weight estimates.

Even though it has a fancy name, the most common type of a neural network model, the multilayer perceptron, can be perceived as an extension of a regression model.

The prediction formula used to predict new cases is like a regression model but with a very flexible addition. This addition enables a trained neural network to model any association between the input variables and the target. Flexibility comes at a price because of the lack of interpretability and the lack of a built-in method for selecting useful inputs.

The neural network model is arranged in layers. The first layer is called in the input layer, consisting of the input variables. The second layer is called the hidden layer, consisting of the hidden units or neurons. It is possible to have multiple hidden layers with multiple hidden units in each hidden layer. The final layer is the target layer, consisting of the response or target.

Figure 4.4: Neural Network Diagram

$$\hat{y} = W_{00} + W_{01}.H_1 + W_{02}.H_2 + W_{03}.H_3$$

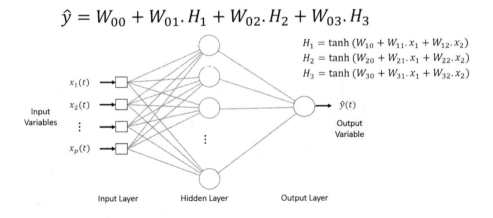

$H_1 = \tanh(W_{10} + W_{11}.x_1 + W_{12}.x_2)$
$H_2 = \tanh(W_{20} + W_{21}.x_1 + W_{22}.x_2)$
$H_3 = \tanh(W_{30} + W_{31}.x_1 + W_{32}.x_2)$

$x_1(t)$

Input Variables $\quad x_2(t)$

$\vdots$

$x_p(t)$

$\hat{y}(t)$

Output Variable

Input Layer   Hidden Layer   Output Layer

Neural networks predict cases using a mathematical equation involving the values of the input variables. The inputs in the input layer are weighted and linearly combined through hidden units or neurons in the hidden layer. The hidden units include a link function, called an activation function in neural networks, to scale the outputs. The result is compared to the observed values in the target layer, the residual is computed, and the weights are re-estimated.

The output from a neural network with one hidden layer is a weighted linear combination of the mathematical functions generated by the hidden units. The weights and biases give these functions their flexibility. Changing the orientation and steepness of these functions, and then combining them, enables the neural network to fit any target.

After the prediction formula is generated, obtaining a prediction is simply a matter of plugging the predictor variable values into the hidden unit expressions. In the same way as regression models, data scientists obtain the prediction estimates using the appropriate link function in the prediction equation.

Neural networks are universal approximators that can model any relationship between the predictor variables and the target. Given the unlimited flexibility of neural networks, they are robust to outliers.

Most predictor variable and target relationships can be modeled with one to two hidden layers. However, the number of hidden units required is a more subjective question to answer. If the neural network has too many hidden units, it will model random variation as well as the desired pattern. If the neural network has too few hidden units, it will fail to capture the underlying signal. Therefore, specifying the correct number of hidden units involves some trial and error. Data scientists can determine the appropriate number of hidden units by using the goodness-of-fit statistic on the validation data set. For example, they can fit several models with a different number of hidden units and choose the model with the lowest goodness-of-fit statistic on the validation data set.

Finding reasonable values for the weights is done by least squares estimation for interval-valued targets and maximum likelihood for categorical targets. The search of the weights involves an optimization process, which applies an error function, or objective function, to update the previous weights at each iteration. When the target variable is binary, the main neural network regression equation receives the same logit link function featured in logistic regression.

Complex optimization algorithms are a critical part of neural network modeling. The method of stopped training starts with a randomly generated set of initial weights. Training proceeds by updating the weights in a manner that improves the value of the selected fit statistic. Each iteration in the optimization process is treated as a separate model. The iteration with the best value (smaller for some statistics, larger for others) of the selected fit statistic on the validation data set is chosen as the final model. To avoid overfitting, a method called weight decay is applied during the search for the right set of weights. It is observed that every time the shape of the activation function gets too steep, the neural network overfits. What causes the shape of the activation function to get too steep is the weights growing too large. The weight decay method penalizes the weights when they grow too large by applying regularization. There are mainly two types of regularizations: Lasso or L1 (penalizes the absolute value of the weight) and Ridge or L2 (penalizes the square of the weight). Another important parameter is the learning rate, which controls how quickly training occurs by updating the set of weights in each iteration.

Neural networks are one of the fastest scoring nonlinear models as they can efficiently score large volumes of data. For example, neural networks are used by banks to assess all their credit card transactions for potential fraud patterns. Neural networks also are commonly used to provide real-time evaluation and monitoring of stock and futures movements.

Neural networks are most appropriate for pure prediction tasks since it is difficult to decipher how the predictor variables affect the target. This is bad for regulated markets such as credit scoring in banks since it will be difficult to explain how the predictor variables affected the predicted target value. One solution is to use decision trees to interpret the neural network's predictions. Simply define the target as the values of the predictions from the neural network. Notice that the decision tree is not explaining the neural network model, but instead it is explaining the results provided by the neural network by predicting its outcomes. In terms of business actions, it can suffice to understand the pattern behind the prediction. In regulatory terms, this approach probably will not suffice.

Another solution is to use model interpretability plots. One example of a model interpretability plot is the partial dependence plot that reveals how the target prediction changes as the values of the inputs are changed. These approaches do not make neural networks directly interpretable, but they allow data scientists to gain insight into them.

Although neural networks are one of the fastest scoring nonlinear models, they take a lot of time when fitting, or training, the model. This is especially true as the number of hidden units and layers increase.

Although neural networks have unlimited flexibility, they sometimes do not perform as well as simpler models such as linear regression. This situation occurs when the signal to noise ratio is low.

When there is a strong pattern, called the signal, relative to the amount of variation, called the noise, the signal to noise ratio is high. In this situation, neural networks usually outperform simpler models such as linear regression. However, when there is a weak pattern relative to the noise, then the signal to noise ratio is low. The linear regression and neural network models might offer a comparable fit to the data as shown in Figure 4.5. Therefore, there would be no advantage to using the neural network model in this situation.

Use Case: Bad Debt

One of the major problems in telecommunications is bad debt, whether it is unintentional or intentional. There are many reasons for a customer to have unintentional bad debt, like unemployment, unexpected expenditures, and so on. There are also many ways companies can handle this bad debt: renegotiating the debt, partially collecting the debt, or partially suspending the services. For intentional bad debt, there is nothing to do, unless it can be prevented. In certain markets, like many in South America, there is also an aggravating circumstance. Companies providing utility services pay sales taxes to the government when they issue the bill. When those customers do not pay their bills, the utility companies do not get the revenue associated with the invoices, but even worse, they do not get back the taxes already paid to the government. It is a double loss.

Figure 4.5: Signal-noise Ratio Effect in Neural Networks

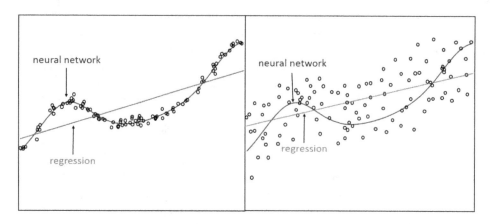

This type of problem can be tackled by using an accurate supervised model, even though it is not possible to interpret it. A neural network can be trained on the historical invoices to capture the relationship of the usage and the intentional bad debt. Thinking about just the top percentage of cases based on the predictive probability, there is a huge opportunity here to save money in terms of taxes. This model runs right before the billing process and creates a list of the invoices most likely not to be paid. Even if the threshold is raised to the top 5% or just the cases where the predictive probability is greater than 92%, the savings in taxes can be substantial. What does the company do in this case? It does not issue the bill. It sounds weird. However, for the top cases in terms of predictive probability, the company does not issue the bill and that can save millions of dollars. Of course, there are always errors in models, such as misclassified cases. Unfortunately, these customers will have their services temporarily disconnected, until they contact the Call Center to complain. This is the moment when the company realizes the misclassification. Fraudsters usually do not complain to the Call Center. When a misclassified customer complains about the service, the company immediately needs to reactivate the

Figure 4.6: Saving in Bad-debt Approach by Using Advanced Analytical Models

Best hit rate	5%
Average billing	$41
Population	480,954
Total billing	$ 19,719,114

Overall hit rate	92%
Correct population	444,882
Billing associated	$19,240,162
Taxes avoided	$5,472,048

Misclassification rate	8%
Incorrect population	36,072
Billing associated	$1,478,952
Taxes	$443,685
Revenue uncharged	$1,035,266

Total taxes avoided	$4,436,782 yearly

services and replace the customer's loss. However, considering all losses and savings, the model still can avoid millions in taxes.

The simulation in Figure 4.6 shows the possible financial return by deploying a bad debt model. For example, in some markets, telecommunications companies need to pay taxes when they issue a customer bill, sometimes around 33%. If customers do not pay their bills, the company cannot recoup the taxes. This is the risk of the business. However, some of the bad debt is fraud. What if the company identifies the fraud before issuing the bill? The damage by the fraud is already done. But the company can at least avoid the taxes. The predictive model evaluates all bills before issuing them. Every bill has an associated likelihood identifying the risk of that bill associated with fraud or not. Considering the top 5% of the bills with higher predictive probability, the overall accuracy is 92%. That means, the model can lead to a wrong decision in 8% of the cases. Considering an average bill of USD 41 and a population of possible intentional bad dept (on the top 5%) of 480,954, the total billing amount is about USD 19.7 million. Making the right decision in 92% of the cases would avoid USD 5.5 million in taxes. Making a wrong decision in 8% of the cases would cost to the company USD 1 million. The savings would still reach over USD 4.5 million at the end. In a real scenario of fraud, all these customers identified as fraudsters would not have their bills issued and would have their services cutoff. The good customers, the 8% included in the model's mistake, would be in touch with the company to determine what happened. The services would be resumed, and the bill would be reissued, reducing the loss caused by the predictive model.

Summary

This chapter introduced some machine learning algorithms for supervised modeling like ensemble of trees (forest and gradient boosting) and artificial neural networks. Situations that could benefit from using machine learning supervised models, some advantages in terms of performance, flexibility and accuracy, and some disadvantages, such as the lack of interpretability for these machine learning techniques were also described. Approaches to interpret the outcomes from supervised machine learning models were described as a method to mitigate the lack of interpretability in such algorithms.

Chapter 5: Advanced Topics in Supervised Models

Chapter Overview

In this chapter, you will learn about advanced topics in machine learning and supervised learning modeling. This chapter focuses on nontraditional machine learning models like support vector machines and factorization machines, as well as advanced methods for supervised modeling, such as ensemble models and two-stage models.

Support vector machines are a very efficient classifier, allowing data scientists to fit a model regardless of the functional form, or the relationship between the input variables and the target. Factorization machines are a common model for recommendation systems, allowing data scientists to predict, for example, customer ratings for items.

The main goals of this chapter are:

- Describe how advanced machine learning and artificial intelligence relate to the field of data science.
- Identify situations that could benefit from using models based on advanced machine learning and artificial intelligence such as support vector machines and factorization machines.
- Describe different types of advanced supervised models that rely on machine learning and artificial intelligence and the circumstances under which they might be used.
- Describe different types of advanced methods to develop supervised models based on ensembles of machine learning models and two-stage models.
- Understand when to use advanced methods of supervised machine learning models like ensemble models and two-stage models and the benefits to use them to solve business problems.
- Interpret the results of advanced supervised models that rely on machine learning and artificial intelligence like support vector machines and factorization machines.
- Interpret the results of advanced methods of supervised models like ensemble models and two-stage models in solving real-world problems.

Advanced Machine Learning Models and Methods

Chapter 5 covers two advanced machine learning models and two advanced methods to train supervised models. The machine learning models that we will discuss are:

- Support Vector Machines – this model implemented in SAS can currently handle binary and continuous targets.
- Factorization Machines – this model is an extension of the linear model and it can handle continuous targets.

The advanced methods that we will discuss are:

- Ensemble Models – this model combines several models to predict the target value. The target can be continuous or discrete.
- Two-Stage Models – this model allows the modeling of a discrete and a continuous target. The model first predicts the binary response and if the prediction is 1, then the model predicts the continuous response.

Support Vector Machines

Support vector machines (SVM) are one of the newest machine learning models presented to solve real-world problems. SVM was created in the 90s, and it is a robust model to classify categorical or continuous targets. Like neural networks, these models tend to be black boxes, but they are very flexible. Support vector machines automatically discover any relationship between the input variables and the target. Data scientists do not need to specify the functional form, or the relationship between the inputs and the target before fitting the model.

Support vector machines were originally developed for pure classification tasks to solve pattern recognition problems. In other words, the model makes decision predictions instead of ranks or estimates. In that way, the SVM separates the outcomes of a binary target into two classes, for example, squares and circles. Support vector machines can now be used for regression tasks as well. In the simple example shown in Figure 5.1, the goal is to classify dark squares versus light circles. There are many classification rules or "regression lines" that can be used to separate the square and circle cases. In fact, if the data is linearly separable, as shown in the figure, there are a limitless number of solutions, or lines, to separate squares and circles or any cases in a binary target. Is there an optimal solution considering all possible lines that split the squares and circles? Given two input variables, the SVM is a line. Given three input variables, the support vector is a plane. With more than three input variables, the support vector is a hyperplane.

For mathematical convenience, the binary target is defined by values +1 and -1, rather than the usual 1 and 0 in logistic regression. Because the linear separator equals 0, classification is determined by a point falling on the positive or negative side of the line. In other words, if the outcome is positive, then the case fits to one class. If the outcome is negative, then the case fits to the other class.

This is a quite simple linear problem to start with. Finding the best solution to a linear classification problem is an optimization problem. The SVM gets more complicated when the problem is not linearly separable. In Figure 5.1, think of the vector W as the mechanism that affects the slope of H (the optimal line that correctly classifies the observations).

The formula for H is shown below. The bias parameter b is the measure of offset of the separating line from the origin, or the plane in three dimensions or hyperplane in higher dimensions. The quantity

$\langle w, x \rangle$ is the dot product between the vectors w and x. A dot product is a way to multiply vectors that result in a scalar, or a single number, as the answer. It is an element-by-element multiplication and then a sum across the products. The algorithm of support vector machines selects values for w and b that define the optimal line that correctly classifies the cases.

$$H = \left\{ \langle w, x \rangle + b = 0 \right\}$$

How are the values of w and b chosen? Support vector machines try to find the decision boundary that maximizes the margin. Margins are the perpendicular distances between the line H and the shapes closest to the line. The SVM is trying to find the hyperplane that maximizes the margin with the condition that both classes are correctly classified.

The properties of the hyperplane that maximizes the margin are described by the support vectors, also called carrying vectors. These are the points closest to the hyperplane and they determine the location of the hyperplane. In other words, they determine the location of H. Because the hyperplane depends only on data points and not predictor variables, the curse of dimensionality is minimized. In Figure 5.2, only five data points are the support vectors that are used to determine the location of the hyperplane, which is based on the parameters of w and b.

Figure 5.1: Linear Separation in Support Vector Machines Training

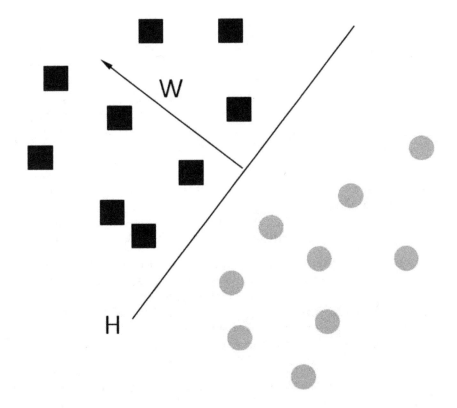

Figure 5.2: Carrying Vectors in Support Vector Machines

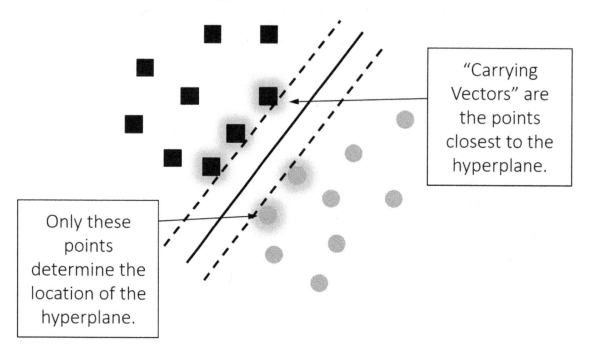

In most realistic scenarios, not only are the data not linearly separable, but a classifier defined by a hyperplane would make too many mistakes to be a viable solution. In that case, one solution would be to apply a polynomial transformation so that the input data would be projected into higher dimensions and then find the maximum margin hyperplane in this higher dimension. For example, the data might not be linearly separable in two dimensions, but it can be separable in three or more dimensions. This can be generalized to higher dimensions. It might be the case that is it not possible to find a hyperplane to linearly separate the classes considering a particular input data, but we could find a hyperplane in a higher dimension that splits the input data when it is set in a higher dimensionality.

The dot product to transform the support vectors in a higher dimensionality, such as polynomials, can be extremely computationally intensive. To mitigate this drawback, the algorithm to calculate the support vectors uses a kernel function, such as a polynomial kernel of degree two or three, to compute the dot product of the two vectors that is much less computationally intensive. Using the polynomial kernel function instead of using the computationally intensive dot product is known as a *kernel trick*.

One of the strengths of SVM is that data scientists do not need to specify the functional form or know the type of the relationship between the input variables and the target. Since there is no functional form, SVM are robust to outliers in the input space. They are also highly effective classifiers when the data points are separable in the input space. Furthermore, SVM are less affected by the curse of dimensionality because it is less affected by the number of predictor variables compared to other models. The hyperplane depends only on data points and not predictor variables.

One of the weaknesses of SVM is that the model itself is a black box with no interpretable parameter estimates or rules and thresholds. Support vector machines work very similarly to neural networks and

Figure 5.3: Increasing Dimensionality to Find a Hyperplane to Linearly Separate Classes

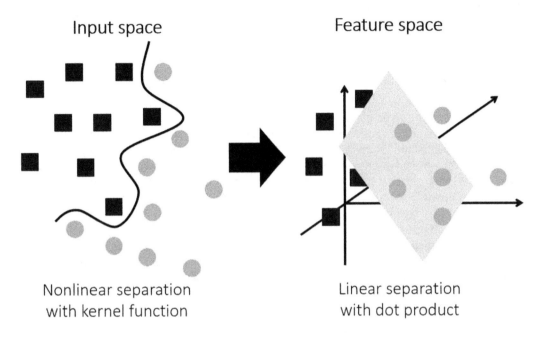

Input space

Feature space

Nonlinear separation
with kernel function

Linear separation
with dot product

tree-based ensemble models like random forest and gradient boosting when the subject is interpreting the results. In regulated markets where the relationships between the predictor variables and the target are important to understand the outcomes, and mostly to explain the results, SVM would have limited usefulness. As was mentioned before in neural network models, a surrogate model, such as a regression or a decision tree, can be used to explain the outcomes from a SVM. However, by using a surrogate model the data scientists are not explaining the models but instead they are interpreting the results in terms of the business actions. Some regulators might not accept this approach.

Use Case: Fraud in Prepaid Subscribers

Support vector machines are useful in pure prediction tasks such as fraud detection. In these cases, a full interpretation of the model outcomes is not necessary, and the main interest is in the classification results or the model decisions. A scenario like this is when a telecommunications company creates a marketing promotion to boost new cell phone operations. The promotion gave subscribers prepaid telephone minutes to make calls at the same amount as the minutes received from any telephone.

Fraudsters immediately identified an opportunity in this scenario. The fraud occurs when the criminals acquire fixed lines to connect to a computer that uses a soft switch that is a call switching node in a telecommunications network. From the soft switch, the criminals generate robocalls to hundreds of cell phones to obtain the prepaid minutes. Of course, the criminals had no intention to also pay for the fixed lines. In this case, it is a double fraud against the telecommunications company. Criminals can also use altered pay phones to generate these calls, and then the prepaid credits, or illegal extensions to switches to do the same, to make calls to prepaid cell phones and generate credits to them. Once these prepaid cell phones are full of credits, the criminals sell these prepaid phones on the black market. In many markets, there is no need to provide any information to the carrier when purchasing prepaid subscriber identification module (SIM cards).

The main goal of the fraud detection model is to detect unusual behavior in receiving calls and block the prepaid minutes credit. Some useful variables include no active behavior in making calls, exceedingly high and consistent behavior in receiving calls, very narrow list of originating numbers, and originating numbers at the same place, sometimes at the same address. The SVM model classifies the cases based on a binary target, 0 or 1, or in this case, fraud or non-fraud. This classification is based on the posterior probability of the event, or the likelihood of a prepaid fraud. Based on the posterior probability, the prepaid phone numbers with the highest posterior probability are sent to a deny list that cancels their prepaid credits. Fraud detection models can save telecommunications companies millions of dollars each month. They should be embedded in transactional systems to timely monitor the events. They must be closely monitored to assess their performance to predict fraud cases. They are often updated frequently, as fraud events change often. For all these reasons, the main goal for fraud detection models is the performance, rather than the interpretability, of the model. This is a perfect case for neural networks, random forests, gradient boosting, and support vector machines.

Factorization Machines

There is a class of business applications that involves estimating how users would rate some items. In this business scenario, very often companies do not have much information about the users nor about the items. The main goal in this type of model is to evaluate relationships between users and items. For example, consider a streaming based company selling on-demand movies to users. This company has little information about its customers (users) except the movies (items) they watch. Users can eventually rate the movies that they watch. If we create a matrix considering users and movies, considering the combination of a rating given by a user to a movie, this final matrix will be very sparse. We can expect, for example, millions of users and hundreds of thousands of movies. Not all customers rate all movies. In fact, few customers rate movies, which leads to a huge matrix comprising a substantial number of missing ratings. In other words, most of the cells in that matrix would be missing. The challenge here is to estimate what would be a user's rating to a particular movie.

Factorization machines are a new and powerful tool for modeling high-dimensional and sparse data. A sparse matrix is the matrix of users and items of the movie company we just mentioned. The main goal of factorization machines is to predict the missing entries in the matrix. By estimating the missing entries, the company would know all ratings for all movies, considering all users. That information would allow the company to find items that users would give high ratings and then recommend those items to each user (except the ones who have rated those items).

Factorization machine models are used in recommender systems where the aim is to predict user ratings on items. There are two major types of recommender systems. One type relies on the content filtering method. Content filtering assumes that a set of information about the items is available for the model estimation. This set of information is often referred to as side information, and it describes the items that will be recommended. By having the side information available, the content filtering has no problem in estimating the rating for a new item.

The second type is based on collaborative filtering. This method does not require additional information about the items. The information needed is the matrix containing the set of users, the set of items, and the ratings of users to items. The combination of all users and items creates a large, sparse matrix containing lots of missing ratings. The collaborative filtering method works well in estimating ratings for all combinations of users and items. The problem with the collaborative filtering method is the inability to estimate added items that have no ratings.

Figure 5.4 shows an example of collaborative filtering. In this case, all the information the company has is the existing ratings, that can be represented in a matrix. Not all the users rate all the items, so some of the ratings will be missing.

Factorization machines estimate the ratings by summing the average rating over all users and items, the average ratings given by a user, the average ratings given to an item, and a pairwise interaction term that accounts for the affinity between a user and an item. The affinity between users and items is modeled as the inner product between two vectors of features: one for the user and another for the item. These features are collectively known as factors.

Matrix factorization techniques are usually effective because they enable data scientists to discover the latent features underlying the interactions between users and items. The basic idea is to discover two or more matrices that when they are multiplied together returns the original matrix.

Figure 5.4: Table of Users and Ratings with Missing User Ratings

	2		1			4				5				?	
	5		4				?		1		3		2	5	
		3		5			2							1	
4			?			5		3		?			?		5
		4		1	3				5			3	4		
			2				1	?			4				2
	3		3		1		5		2		1			1	
		3				1			2		3				?
?		4			5	1			3				2	4	
			3				3	?				5	?		2

Imagine the following scenario as shown in Figure 5.5. There is a set **U** users and a set **I** items. The matrix **R** of size |**U**| **X** |**I**| contains all ratings that users have assigned to the items. The number of **K** factors can be arbitrarily defined. These **K** factors represent the latent features about the relationship between the set **U** of users and the set **I** of items. The main goal is to find the two matrices **P(a|U|×K)** and **Q(a|I|×K)** where the product of these two matrices approximates **R**: **R ~ P×Q^T=R**.

Each row of **P** represents the strength of the associations between a user **u** and the features **k**. Similarly, each row of **Q** represents the strength of the associations between an item **i** and the features **k**. To get the prediction of a rating of an item i_j by a user u_i, we need to calculate the dot product of the two vectors corresponding to u_i and i_j: $\hat{r}_{ij} = p_i^T q_j = \sum_{k=1}^{k} p_{ik} q_{kj}$

One method to obtain the matrices **P** and **Q** is by first initializing these two matrices with some values. These values are exactly the existing ratings (users rating items). We calculate how different their product is to **R**, and try to minimize this difference iteratively. This method is known as gradient descent, and it aims to find a local minimum of the difference.

The difference is the error between the estimated rating and the observed rating, and it can be calculated by the following equation for each user-item pair: $e_{ij}^2 = \left(r_{ij} - \hat{r}_{ij}\right)^2$

The squared error is considered because the estimated rating can be either higher or lower than the observed rating.

Use Case: Recommender Systems Based on Customer Ratings in Retail

A recurrent business problem in many industries is recommending the right product or service to the right customer. Imagine an online sports retail company that sells many different products for several types of sports. For example, the sports might include soccer, baseball, basketball, and cycling, and the products might include clothes and equipment.

Figure 5.5: Matrix Factorization to Estimate the Missing Values

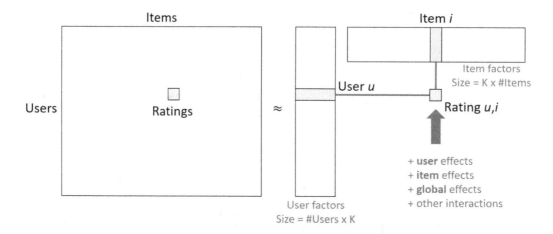

Imagine also that this company has a reward program that incentivizes customers to sign in when shopping or browsing. When customers sign-in, they can be identified, and all transactions over time can be matched. Customers are asked to rate the products that they purchased after a certain period, allowing them to experiment with the product and then provide a more accurate rating.

A recurrent promotion can identify customers who purchased similar products and have similar ratings. This combination of customers and items creates a huge matrix. A factorization machine model estimates all missing ratings, considering customers who do not have the rated products. A reasonable scenario considers customers who did purchase a product but did not rate it. A target marketing campaign selects all products with high estimated ratings for customers. Each customer receives a promotion for each one of the products with high estimated ratings. Before launching the campaign, a filter is applied to remove all items already purchased by the customers. As customers can purchase products and not rate them, an estimated rating is computed for those customers. By removing the products already purchased by the customer, the company can avoid sending out messages and promotions about products the customers already have.

Ensemble Models

Ensemble models create new models by combining the predictions from multiple models consisting of different modeling techniques or different modeling structures (multiple neural networks with different numbers of hidden layers and hidden units). The combined model is then used to score new data. For binary targets, data scientists can take the mean of the posterior probabilities as the final predicted probability. Another approach is to derive a decision (for each observation scored in the model, if the posterior probability is greater than 50% then class 1, otherwise class 0) and see how many models predict a 1 versus a 0. Then take a plurality vote of the predicted class. For continuous targets, the

Figure 5.6: Ensemble Models

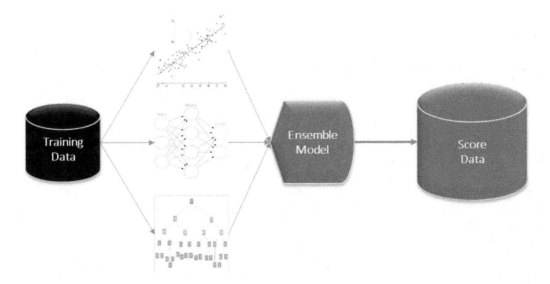

usual approach is to make the final estimation as the mean of the estimations for each model in the ensemble structure.

The commonly observed advantage of ensemble models is that the combined model is usually better than the individual models that compose it. "Better" has two meanings—better in terms of generalization because the final model accounts for relationships discovered from distinct models. The combination of these models can better account for the variability of future data than an individual model. The ensemble model can also be better in terms of accuracy. The overall estimation can present less error than individual models, or the overall decision can present less misclassification error than individual models. It is important to notice that the ensemble model can be more accurate than the individual models if the individual models disagree with one another. Data scientists should always compare the model performance of the ensemble model with the individual models and evaluate when to deploy the ensemble model or one of the individual models from the ensemble structure.

When we think about the benefits of ensembles of decision trees such as random forests and gradient boosting, we notice the same benefits of ensemble models. These models can have low bias and low variance and be more robust to overfitting. The ensemble model can generalize better to future data. And the ensemble model can be more accurate. However, like gradient boosting and forest models, ensemble models lack interpretable parameter estimates. Therefore, if understanding the relationships between the input variables and the target is a priority, ensemble models are not useful. The ensemble models might also be slow in scoring new observations since many models are in production.

If the relationships between the input variables and the target are accurately captured by a single model, the ensemble approach offers no benefit. First, ensemble models are more computationally expensive since several models are being fitted and the results are averaged. Second, the ensemble model cannot be interpreted. This is particularly important if the single model is an interpretable model like a regression or a decision tree.

Use Case Study: Churn Model for Telecommunications

An example of where ensemble models might be especially useful is in churn in telecommunications. Telecommunications companies usually have millions of customers, and customers might present several distinct types of behavior of usage, payment, and even making churn. Eventually, a single model would not be able to capture all of these different relationships between the input variable and the target for subgroups of customers. Of course, one option to mitigate this is to first create a customer segmentation (clustering) and then create one churn model for each cluster. However, even inside the same cluster, we can observe different customer behaviors. An ensemble model can work well in capturing all these different behaviors and correlating them to the target, in this case the churn event.

Even though we have the ensemble model as the final model, we also have the individual models that contributed to the ensemble approach. As mentioned before, the ensemble model lacks interpretability. However, data scientists can use the ensemble model as the decision in a retention campaign (offer something to the customer or not) and can also use some of the individual models (the one with the highest predictive probability) to eventually explain the result or a characteristic associated to the churn event. Data scientists need to understand that business departments require more than just a probability. They also want to understand the reasons customers are making churn, the reasons customers are not going to pay their bills, and the reasons customers are going to purchase a product. All these reasons help business departments create the marketing campaigns, the offerings, the marketing messages, and the channels to contact the customers. A minimum of interpretability is needed to support a full business action. Therefore, the combination of the ensemble model to select the customers and the individual models to interpret the results can extract most of the benefits from analytical models.

A churn model quite often uses all the information about the customer, considering the time frame perspective. How the customers are changing their behavior right before the churn, and how these characteristics are correlated to the target, or in this case the churn event. As shown in Figure 5.7, common input variables in churn models include:

- Demographic variables: customer characteristics such as age, gender, and education
- Usage variables: how much usage of products and services and indicators over time to characterize the trends of usage
- Product/service portfolio: what bundles and packages the customer has and for how long
- Revenue behavior: how much revenue the customer generates, for how long, and indicators over time to characterize trends about the average revenue per user
- Payment behavior: how many defaults the customer has had in the past and the length of the delay
- Contract information: how much time before the contract ends
- Contacts in the call center: the frequency and recency of contacts to the call center, including the reasons and the resolutions
- Defects reported: the frequency and recency of defects reported by the customer, the types, the time to fix them, and the resolution approach

Figure 5.7: Ensemble Model to Classify Churn in Telecommunications

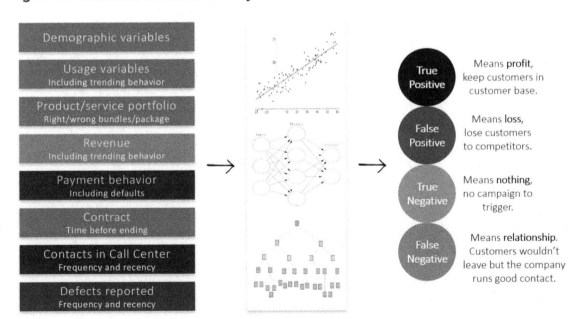

Figure 5.8: Two-stage Models

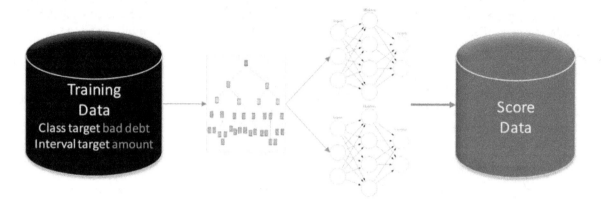

The predicted probability from the final model can be used to decide which customers receive a promotion campaign, most likely offering something to retain them in a long-term perspective.

Two-stage Models

Two-stage models enable data scientists to model a class target and then use the predicted probability, or the decision, to split the observations and feed them into another model, normally to train a model with a continuous target. The two models can use different algorithms and different predictor variables. Therefore, the first model classifies a binary target. Based on the decision of the first model (0 or 1), another model is trained to estimate a continuous value (just for the decisions of 1). A good example is to model the probability of unmanageable debt, and if the predicted class is 1, then model the amount of unmanageable debt.

Two-stage models provide data scientists the probability of the event and the predicted amount. For example, if the data scientists are modeling a fraud event, they can obtain the probability of the transaction being fraudulent in the first model, and then, for all those transactions classified as fraudulent, they can estimate the amount of fraud involved in the transaction in the second model. These two pieces of information can be used to determine the amount of fraud, rather than just the probability of the fraudulent event.

If the predicted probability is accurately estimated in the first stage of the model, then there might be an increase in the accuracy of the predicted amounts in the second stage. For example, in two-stage default models, the accuracy of the predicted probability of default could increase the accuracy of the predicted amount of default.

Data scientists need to be careful about which models are used in the two-stage model because a poorly specified model in either stage impacts the accuracy of the predictions, and therefore the estimations. The two targets also need to be correlated, or else two independent models would be sufficient.

Use Case: Anti-attrition

A good business example for two-stage models is problems related to customer attrition. When a customer calls the company call center with a complaint, a predictive model can be fit to predict what the customer will do if the company does not rectify the complaint. For example, if the customer feels they have been charged an excess amount, will the customer go to a consumer protection agency, or to a regulatory agency, or go straight to court if their complaint is not rectified? Each one of these jurisdictions has its own cost for the company as shown in Figure 5.9. If the problem can be rectified at the call center level, it is cheaper for the company. If the problem reaches the consumer protection agency, the cost for the company is higher. If the problem reaches the regulatory agency, the cost can be even higher as some penalties can be added. However, if the problem goes to court, the cost to the company can be enormous because in addition to penalties, there might be some civil reparations involved.

To accommodate complaints and optimize the way the company handles reducing the costs associated to higher complaint escalation, a first model will classify what action each customer is likely to take if their complaint is not remedied by the call center representative. A second model estimates the loss associated to the case if the customer complaint is not rectified and reaches a higher level for resolution.

A decision tree is used to classify the probability of each customer complaining to a higher authority. For example, what is the probability the customer escalates the problem to the consumer protection agency, the probability the customer escalates the problem to the regulator agency, and the probability the customer escalates the problem to the court? For example, assume that the highest probability of action for customer A is to escalate the problem to court. This probability was produced by a decision tree. A neural network is then used to estimate the loss for the company if customer A wins the case in court. Notice in Figure 5.10 that the first model, the decision tree, is interpretable and helps the business department understand the reasons associated with the customer complaint and escalation approach. The second model, the neural network, is not interpretable, but at this point the company is much more interested in understanding the amount of loss associated with the customer action.

By having all this information, the call center representative knows the probability of the customer escalating the problem to a higher authority and the estimated company cost of this action. The call center representative can have multiple messages and approaches depending on the combination of authority and loss associated to each customer. Suppose that for customer A, the representative

Figure 5.9: Complaint Workflow in Telecommunications

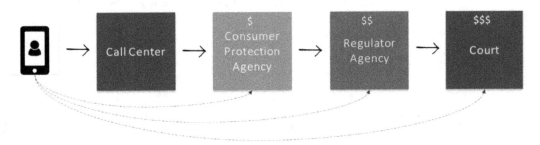

Figure 5.10: Two-stage Model to Classify Possible Complaints and Estimate Associated Losses

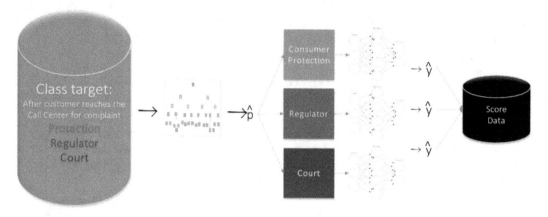

realizes how likely the customer is to escalate the problem to court, and the high loss associated to it, and then can decide to suspend the invoice until the case is fully analyzed. If the customer is correct about the overcharge, the company can easily remove that specific invoice and provide to the customer a pleasant experience. If the customer is wrong about the overcharge, the company can reissue the invoice and charge the customer for that usage.

Summary

This chapter introduced more advanced techniques for supervised models, including support vector machines and factorization machines. The first model is a highly effective classifier, and the second one is used in recommender systems. The recommendations are based on ratings customers give to specific products or services. If the ratings are high, those products or services are recommended to the customers. If the ratings are low, the company should not recommend these products or services to those customers. Both techniques are less common in the data science field and therefore less often found in business applications. However, support vector machines and factorization machines are very promising techniques to solve real-world business problems.

Other topics covered in this chapter are ensemble models and two-stage models. Both analytical approaches are very effective in solving business problems. Ensemble models usually provide a more generalized model, avoiding misclassified predictions in sensitive events such as fraud, diagnosis, and insolvency. Two-stage models are good approaches when the need is to classify an event and then to estimate a value. For example, classifying the likelihood that an event is fraudulent and then, if the probability is high, how much the fraud would cost.

Additional Reading

1. Abbey, R., He, T. and Wang, T. 2017. "Methods of Multinomial Classification Using Support Vector Machines." *Proceedings of the 2017 Annual SAS Users Group International Conference*. Cary, NC: SAS Institute Inc.

2. Silva, J., and Wright, R. E. 2017. "Factorization Machines: A New Tool for Sparse Data". *Proceedings of the 2017 Annual SAS Users Group International Conference*. Cary, NC: SAS Institute Inc.

3. Gunes, F., Wolfinger, R. and Tan, P. 2017 "Stacked Ensemble Models for Improved Prediction Accuracy". *Proceedings of the 2017 Annual SAS Users Group International Conference*. Cary, NC: SAS Institute Inc.

4. Shirodkar, G., Chaudhari, A., and Chakraborty, G. 2015. "Improving the performance of two stage model using Association node of SAS Enterprise Miner 12.3". *Proceedings of the 2015 Annual SAS Users Group International Conference*. Cary, NC: SAS Institute Inc.

Chapter 6: Unsupervised Models–Structured Data

Chapter Overview

In this chapter, we look at unsupervised models. Unsupervised models are characterized by not having a target. As we learned in Chapters 3, 4, and 5, supervised models consider past events, or historical information, and a target, an outcome of interest that we previously know the results. Unsupervised models, on the other hand, have no target. The main goal is to find insights or to understand trends based on explanatory variables and past data. Past data can be assigned to past events, but those events will not be used as a target to train a model.

Examples of unsupervised models are clustering, association rules, link analysis, path analysis, network analysis, and text mining. This chapter addresses clustering, a method to group observations based on the similarity among observations. Chapter 7 covers association rules. Methods to understand events that take place together include market basket analysis; sequence analysis to identify sequences of steps that commonly occur; link analysis to identify events that are somehow correlated to each other; and text analytics, to analyze text, or natural language, and extract themes, topics, and insights from it. Finally, Chapter 8 covers network analysis and network optimization. These techniques analyze connected events or entities and use algorithms to understand and optimize networks.

In this chapter, we discuss three types of clustering techniques: hierarchical, centroid (also known as partitioning), and self-organizing maps (which is a model-based approach to clustering). The best use of these techniques depends on the business goals and what needs to be achieved by the model and how the results are used in terms of business actions.

The main goals of this chapter are to do the following:

- Explain the purpose of using statistical unsupervised models.
- Understand the concepts associated with clustering.
- Describe diverse types of statistical unsupervised models and the circumstances under which they might be used.
- Describe the steps required to train an unsupervised model using the clustering methods described.
- Interpret the results of a clustering model in terms of business goals.

Clustering

Unsupervised models do not have a target variable. Therefore, the aim is to search for an unknown pattern in the data set. When a target is not present in a data set, the algorithm used to build analytical models is called unsupervised. A common type of unsupervised model is clustering, which attempts to group observations or cases within a data set based on the similarities of the input variables.

Data scientists can use unsupervised models in many business applications, including:

- **Data reduction** is a method to explore patterns in the data set to create a more compact representation of the original information. This is different from most clustering techniques, which seek to group observations. Instead, these methods seek to reduce variables. Though vastly broader in scope, data reduction includes analytic methods such as variable clustering. These variable clusters explain a high percentage of the variability of all original attributes comprised in the cluster. Data scientists can then either use the variable clusters instead of all original variables or select a best representative variable from each variable cluster.
- **Anomaly detection** is a novelty detection method to seek unique or previously unobserved data patterns. These methods find applications in business, science, and engineering. Business applications include fraud detection, warranty claims analysis, and general business process monitoring. Small clusters or quite diverse groups can be considered as an anomaly and be further investigated by analysts seeking unexpected events or transactions.
- **Profiling** is widely used in cluster analysis. The idea is to create rules that isolate clusters or segments, often based on demographic or behavioral measurements. Data scientists might develop profiles of a customer database to describe the consumers of a company's products and services.

The main objective of most clustering methods is to partition the data into groups so that the observations within each group are as similar as possible to each other. The groupings created from the clustering algorithms are also known as clusters or segments. The process to create the groups or segments is a training process. The algorithm can be applied to other data sets to classify new cases. This process is like supervised model scoring. Suppose that a cluster is created based on a set of rules and thresholds upon the input variables. If a new observation follows the same rules and thresholds, we can consider this new case as part of that cluster.

The purpose of clustering is often descriptive. For example, segmenting existing customers into groups and associating a distinct profile with each group might help future marketing strategies and campaign actions. Clustering is also useful as a pre-step for predictive modeling. For example, customers can be clustered into more homogenous groups based on their behavior in using products and services. Then, a predictive model can be trained upon each cluster to predict whether customers make churn or not. Suppose that in a communications company, there are many diverse types of customers, based on the products that they purchase and how they consume these services. However, all of them are susceptible to make churn. As they have quite different behaviors, a single model can be difficult to capture the relationship between the input variables and the target.

However, if we create groups of customers with more similar behaviors, a model for each group captures these relationships.

Clustering is typically done on numeric variables. If there are categorical variables, which is very understandable, they must be represented numerically. Sometimes, a data scientist creates their own dummy codes to represent categories. A *dummy code* means coding a categorical variable into dichotomous variables such as binary 0-1 variables. For example, suppose an input variable has three classes, A, B, and C. The dummy code method creates three variables to represent A as 0 or 1, B as 0 or 1, and C as 0 or 1. While some like the flexibility of creating their own dummy codes, most statistical software can perform this task automatically.

Figure 6.1 shows a data set that contains only two numeric variables: salary and age. These variables are used to create clusters. It seems we can find three distinct groups of observations, or cases, based on the cluster demographics. In other words, if we plot all cases based on their salary and age, we can visually identify that there are three distinct groups of customers.

Clustering methods are highly dependent on units of measurement. In general, variables with large variances tend to have more influence on cluster formation than variables with small variances. For example, in a clustering based on salary and age, we can assume that the variation in all salaries can be greater than the variation in all ages.

Therefore, it is important to have all input variables in the same scale. For example, considering all salaries in a data set, we might observe a range from $10K to $1,000K. Looking at the age, we might observe a range from 21 to 79. Not only is the variation in salary greater than age, but also the scale of salary is much greater than age. A good approach is to standardize all input variables before using them in clustering so that the salary varies from 0 to 1 and the age varies from 0 to 1 as well.

Figure 6.1: Clustering Method to Group by Similar Observations

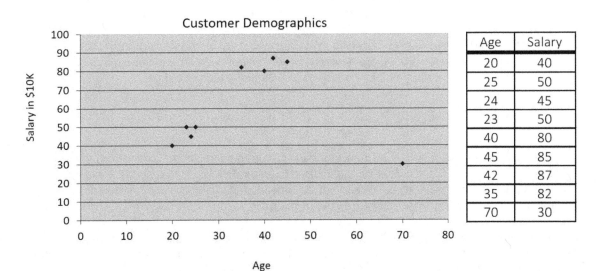

Age	Salary
20	40
25	50
24	45
23	50
40	80
45	85
42	87
35	82
70	30

By doing this, we are giving the same chance for each variable to discriminate the observations into distinct groups.

It is also recommended to use only the relevant information about the customers that is pertinent to the business. For many companies, it is common to maintain a lot of information about the customers and much of it is external and not very related to the business itself.

Categorical variables can be used if they are dummy coded. However, if these variables have many levels, the dummy code generates a substantial number of new variables that are not useful for the clustering technique. As stated before, clustering is typically done on few numeric inputs. Too many variables reduce the overall variability of the data and make it hard for the clustering algorithm to capture the similarity among the observations. Finally, it is good to have independent variables assigned to the clustering process. Different input variables containing the same level of information about the observations increases the dimensionality of the input space and reduces the overall variability.

In summary, some important requirements for clustering are:

- Scale or normalize all attributes.
- Consider relevant attributes on the business goal.
- Do not have too many levels in the categorical attributes.
- Have an interval level of measurement for the attributes (dummy code the categorical attributes).
- Do not have too many attributes.
- Use only independent attributes.

In this chapter, we will learn about the following clustering methods:

- **Hierarchical clustering** creates clusters that are hierarchically nested within clusters at earlier iterations. Agglomerative clustering starts with one cluster per point, and repeatedly merges nearby clusters.
- **Centroid-based clustering** divides a data set into clusters by trying to minimize some specified error function. K-means clustering starts with k clusters and assigns the data points to the nearest center. The algorithm shifts centroids and points over time until no more moves are taken.
- **Self-organizing maps** assign observations to clusters based on the similarities of their attributes. Self-organizing maps (SOMs) are neural networks that provide a topological mapping from the input space to the clusters. Every observation assigned affects the cluster.

Hierarchical Clustering

Hierarchical clustering is a recursive sequence of partitions in the data set of observations. If the data is hierarchically structured, the clustering can be used to generate topic hierarchies. For example, web directories provide a hierarchical classification of documents. Data scientists can restrict the search for documents to only a subset of web pages related to the topic.

The hierarchical clustering algorithms creates a hierarchical structure of clusters, where each level defines a partition of the data set. Hierarchical clustering comes in two forms: agglomerative, which combines the cases iteratively, and divisive, which divides the cases iteratively.

Figure 6.2: Hierarchical Clustering

Agglomerative

1. Assign each observation to its own cluster.
2. Merge the two clusters that are most similar.
3. Repeat step 2 until there is only a single cluster.

Divisive

1. Assign all the observations to a single cluster.
2. Partition the observations that are least similar into two clusters.
3. Repeat step 2 until there is only a single observation in each cluster.

Figure 6.3: Agglomerative Algorithm

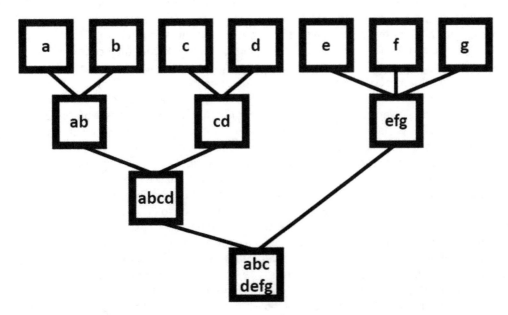

Figure 6.4: Divisive Algorithm

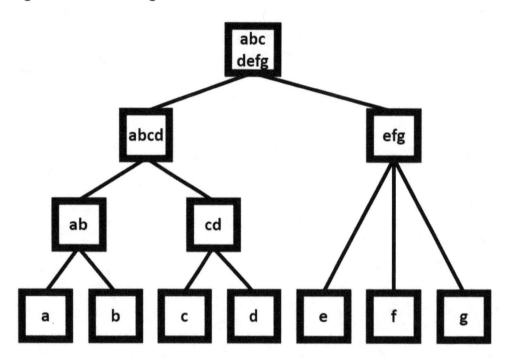

The hierarchical cluster structure is represented by a graph called the dendrogram. A *dendrogram* is a structured graph showing the sequence of divisions in the original data set performed by the hierarchical clustering algorithm. (See Figure 6.5.) The forks represent all the fusions or joining of groups along the clustering process and help data scientists to determine whether there is a natural clustering set throughout that iterative sequence that better groups observations together. In a vertical dendrogram, the fusion level is represented as the height of the fork from the horizontal axis. Short fusion levels indicate fusions of similar clusters. Long fusion levels indicate fusions among heterogeneous clusters, or groups of observations that should not be joined.

Choosing the right number of clusters is not an easy task and depends on the business requirements. There are a few techniques to evaluate and estimate a suitable number for the clusters, including the gap statistic method and domain knowledge. Most techniques investigate statistical analysis to evaluate the level of variability within and without the clusters. The gap statistic method works by evaluating the intra-cluster variation for different values of clusters with their expected values under the assumption of uniformly distributed data. The second and the best approach in most cases is domain knowledge. Subject matter experts can work along data scientists to understand the business requirements, the purpose of the clustering, the practical actions to be taken based on the clustering results and together estimate the right number of groups to support a specific demand. Imagine a company trying to run different campaigns based on different customer behaviors. Even if the gap statistic method tells us that the best number of clusters is 17, it is difficult for the marketing and sales departments to create 17 different campaigns, 17 different promotions, 17 different scripts in the call center, and so on. A business decision can be to establish that four or five clusters suffice for this business purpose.

Figure 6.5: Dendogram

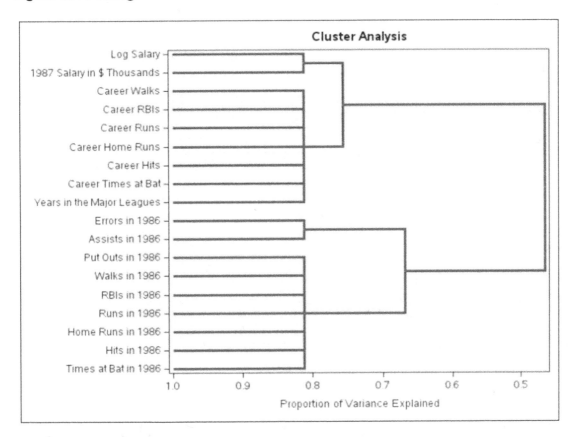

There are three main advantages to using hierarchical clustering.

- Hierarchical clustering is easy to implement and visually interpretable.
- The dendrogram produced is very useful in understanding the sequence of the hierarchical clustering algorithm and helping the data scientists in selecting the right number of clusters.
- Data scientists do not need to specify the number of clusters required for the algorithm. They can interactively investigate the best number of clusters looking at the dendrogram, evaluating the variability explained at each level, and comparing it to the business requirements.

The weaknesses of hierarchical clustering are:

- Hierarchical methods do not scale well to large data sets. Most hierarchical methods increase in processing time as a function of the square, or even the cube, of the number of observations. As the number of variables added to the algorithm increases, the time to run the clustering method also increases. If the data scientist tries to keep the input set small, eventually there is not rich enough information to cluster observations into meaningful groups.

- Previous merges or divisions are irrevocable. For example, if two observations are clustered in step 1, the algorithm cannot undo that connection even if it turned out to be a poor one at the end. Other methods can recursively search for a better distribution, even in further steps.
- Data scientists cannot score new observations using hierarchical clustering. The algorithm is mainly used for descriptive purposes.

Use Case: Product Segmentation

Hierarchical clustering methods can be used in product segmentation. The main goal is to group products and services into similar groups. These diverse groups of products and services can then be used to create distinct packages and bundles to boost cross-sell and up-sell opportunities. Bundles are usually priced attractively to customers and at the same time offer a better option in terms of products and services. In the long term, it retains customers and improves the relationship with them. Bundles and packages more often deliver price discounts but increase time contracts, guaranteeing a solid cash flow for the company.

For example, data scientists in a telecommunications company can use hierarchical clustering to identify groups based on the different products and services the customers use and based on distinct customer behaviors. They can use these clusters to create distinct packages and use them in multiple campaigns to reach out to the customers. These bundles are expected to improve customer satisfaction, contract time, and most important, average revenue per user (ARPU).

Figure 6.6: Cluster Descriptions from Hierarchical Clustering for Product Segmentation

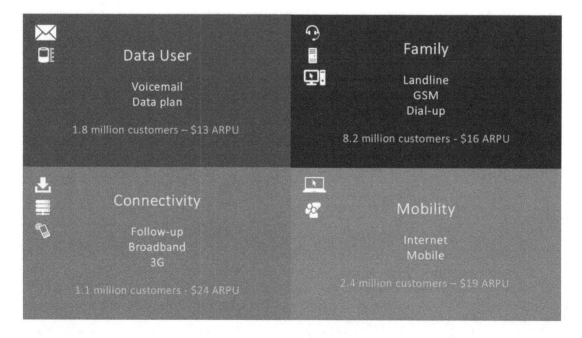

Centroid-based Clustering (k-means Clustering)

The second type of clustering method covered in this chapter is centroid-based clustering. The centroid-based technique is a partitive clustering, which divides the data set into clusters by trying to minimize some specified error or objective function. K-means clustering is one of the most popular partitive clustering algorithms. K-means attempts to divide the observations from a data set by minimizing the Euclidian distance between the observations and the centroids for each cluster. The algorithm consists of four major steps:

1. Choose k initial means as starting "seed" values for the cluster centers.
2. Assign each observation to the cluster center with the shortest distance.
3. Based on the observations assigned, compute a new mean for each cluster.
4. Iterate until the k means stabilize.

The first step in the k-means algorithm is to choose a value for k, or the number of clusters. K represents the objective number of clusters. The term objective is used here because during the clustering process the number of centroids can be excessive to divide all the observations into k clusters. Imagine you have a data set with 100 observations and set k as 102. There are more centroids than observations so that at least two centroids are void. Based on the variance of the inputs for all observations, in some instances some of the centroids defined are void. For example, suppose that all observations have the same values. All the observations will then fall into the same centroid. Of course, both cases are the extreme cases, but it illustrates that k is the goal, the desired number of clusters. We can think of k as the maximum number of clusters.

These centroids, known as seeds, are randomly selected as a first guess of the means of the final clusters. These seeds are typically selected from the sample data, although they can be user specified.

The Euclidean distance from each case in the training data to each seed, centroid, or cluster center is then calculated in this first step. The Euclidean distance gives the linear distance between any two points in n-dimensional space. Cases are assigned to the closest centroid. Because the distance metric is Euclidean, it is important for the inputs to have compatible measurement scales. As in the hierarchical clustering method, all inputs should be numeric, independent, relevant, and standardized (same scale).

Once all observations are assigned to the centroids, the centroids are then updated to equal the average of the cases assigned to the cluster in the previous step. In other words, the centroid should represent the center of all observations. Then, the centroid positions are updated to satisfy this rule.

Now, as the centroids update their positions, eventually some of the observations might be closer to a centroid other than the one to which it was originally assigned. Based on that, the algorithm reassigns all the points to the cluster with the closest centroid.

As observations are reassigned, the centroids need to be repositioned. As the centroids are repositioned, the observations need to be reassigned and the process keeps going until there is no more reassignments for observations.

In Figure 6.7, we observe that three centroids are randomly set. All observations are then assigned to the closest centroid. The centroids are repositioned. Then there are three observations that changed clusters. The mean calculation for the clusters is repeated and the centroids are repositioned. The cluster centers are now the cluster means. The process is continuous until there are no more observations reassigned. To avoid the indefinite running of the algorithm, the member assignment and mean calculation process is repeated until either no significant change occurs in the position on the cluster means, or until the specified maximum number of iterations has been reached. When the process stops, final cluster assignments are made. Each case is assigned to a unique cluster. The cluster definitions can be applied to new cases outside of the training data so that new observations can be assigned to the final clusters.

One of the most important characteristics of k-means clustering is that the algorithm is very simple to implement, can handle large data sets, is quite easy to interpret, and probably the most important, can be used to score new cases or observations.

Even though k-means clustering is probably one of the most popular clustering algorithms in data science, there are some shortcomings. The algorithm requires a guess for the number of clusters present in the data set. Most of the time it is very hard to guess the best number of clusters. Normally, several attempts are made considering multiple numbers of final clusters to decide about

Figure 6.7: K-means Clustering

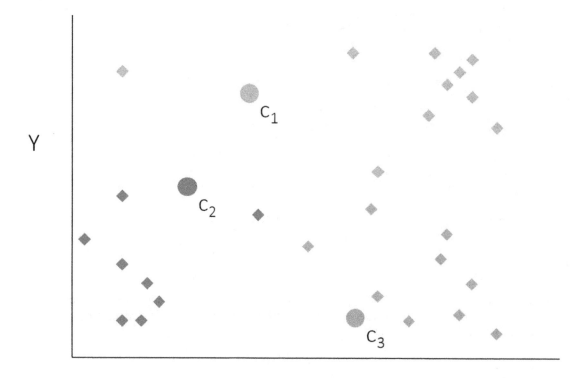

the optimal number of groups. It also depends on the business requirements and the practical actions that will be taken based on the final clustering process. The k-means algorithm is strongly influenced by the initial location of the seeds, by outliers, and even by the order in which the observations are read during the training. Finally, it is very hard to make assumptions about the shape of the clusters. The assumption of the shape of the clusters is that they are spherical. If the true underlying clusters in your data are not spherical, then the k-means algorithm might produce poor clusters.

Use Case: Customer Segmentation

K-means clustering can be used to group customers based on their main characteristics. K-means clustering is very useful when there are many different characteristics about the observations. Some of the attributes include how customers consume products and services, others include their payment history, how they pay, how they delay in payments, and so on. Some customers' attributes are about the revenue that they generate along the way, the profit, or the cost. Some are just about demographic information. All this information together can discriminate groups of customers with similar characteristics and behaviors, and this can be very helpful in marketing campaigns, sales approaches, and bundle and packages experimentation.

For example, if one group of customers shows high usage of products and services, high revenue, and no delays in payments, then it is in the interest of the company to keep these customers happy. A loyalty promotion campaign could be targeted at those customers to guarantee they keep consuming company products and services. Some other customers might match to a different behavior, with average usage and revenue generated, but no payment delay at all. These customers can represent most of the customer base. Even though they are not high revenue generators, as they do not delay their payments and consistently use the products and services, they can be considered as great

Figure 6.8: K-means Clustering for Customer Segmentation

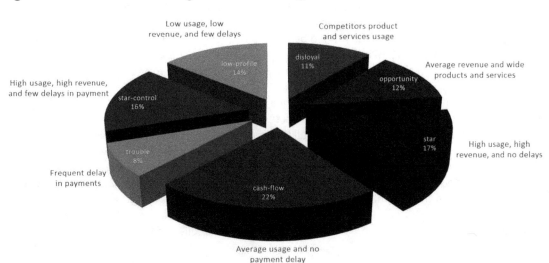

contributors to the company cash flow. They should be effectively monitored and retained as much as possible.

However, if another group of customers shows a frequent delay in payments, eventually with high average bills, then it is in the interest of the company to closely monitor those customers because they are at risk of defaulting on their payments and affect the company operations. Each cluster, or group of customers, holds a set of characteristics that put them together in the same group. This information can be effectively used to better target the customer base in more specific campaigns.

Self-organizing Maps

The self-organizing map (SOM) algorithm is a method that combines the main objectives of a clustering technique and projection algorithm to reduce dimensionality. Not only can this be used to organize or assign observations into clusters, but it can also be used to project multidimensional data onto a two-dimensional feature map, where geometric relationships between the input variables indicate the observations similarity.

Self-organizing maps are also a useful data mining and visualization tool for large and complex data sets. Because they represent a multidimensional input space (multiple input variables) into a two-dimensional matrix (the lattice matrix representing the clusters), they can be used to plot the observations into a two-dimensional chart. This makes it easier to assess input variables distributions across all observations in the data set.

SOM are a type of artificial neural network that is trained using unsupervised learning. They are different from the neural network model that we described in Chapter 4 that was trained based on the target. The SOM algorithm is a neural network that trains upon competitive learning. The idea was first introduced by C. von der Malsburg in 1973 and then developed and refined by T. Kohonen in 1982. Think of it as a neural network with a single computational layer of neurons arranged in rows and columns. Each neuron contains a weight vector with the input variable values and the geometric location in the grid. Furthermore, each neuron is fully connected to the input layer. The model creates a map, or a low dimensional matrix, containing a discretized representation of the input space. The traditional supervised neural network is trained based on an error function. The unsupervised neural network in this clustering perspective is trained based on a neighborhood function to preserve the properties of the input variables. This is the competitive learning assigned to the self-organizing maps.

The low dimensional matrix needs to be set in advance like the k in the k-means clustering. That creates the map for the competitive learning. For example, a two-dimensional matrix is defined to represent all the observations from the input data set and all the input variables. Think about this process as a dimension reduction, as you might have thousands of rows (observations) and hundreds of columns (inputs) being mapped to a smaller matrix, perhaps with three lines and three columns (nine cells or nine groups or clusters). The map is initialized with random weights that represent the input space. In other words, each cell in the map has initial weights to represent the observations inputs.

The neurons' responses are compared. The winning neuron is then selected from the matrix.

Figure 6.9: Kohonen Self-organizing Maps

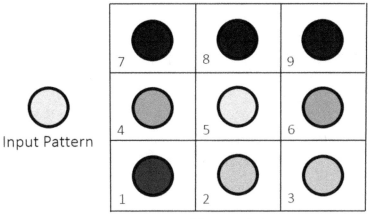

Weights represented by the darkness of balls

Neuron 5 is the most *similar* to the input. Then the *winner* neural is neuron 5.

Input Pattern

Neuron Layer

The next step in the self-organizing algorithm is to randomly select an input vector from the set of training data. Every neuron is examined to determine which weights are most like the input vector. The algorithm used to determine the most similar neuron is a Euclidean distance calculation. The winning node is commonly known as the Best Matching Unit (BMU). The neurons closest to the BMU are called neighbors. Every time an observation is taken for the model training, it is then compared to all cells in the map. In the example in Figure 6.9, we have nine cells to compare. A comparison is made between the input vector and the weights in all cells. The observation is then assigned to the closest cell in the

Figure 6.10: Weights Adjustment Based on Observations Similarity

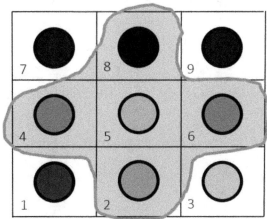

Neuron 5 neighbors are found.

Neurons 2, 4, 6, and 8 have adjusted weights.

Figure 6.11: Map Reconfiguration Based on Weights Adjustment

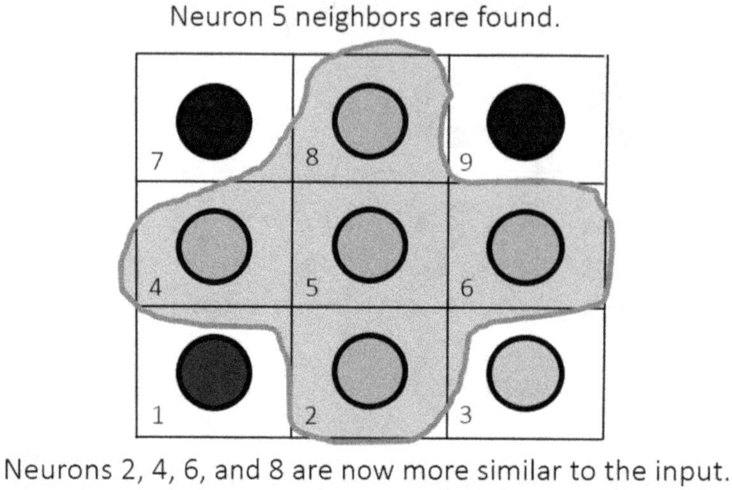

Neuron 5 neighbors are found.

Neurons 2, 4, 6, and 8 are now more similar to the input.

map. The recursive process guarantees that every observation is taken by the model during the training process to compare the input vectors to the cells' weights.

The selected neuron is activated together with its neighborhood.

The weights of the BMU and neurons close to it in the SOM grid are adjusted toward the input vector. The effect of this weight update is to move the BMU and the neighbors closer to the input vector.

In summary, once an observation is assigned to a particular cell, all neighboring cells are adjusted to account for the neighbor's weights. This makes similar observations fall into the same clusters and makes similar clusters get close to each other. This process occurs continuously until all observations in the training samples are taken for the model training and all cells in the map are weight adjusted according to the observations' assignments.

The adaptive process changes all the weights to resemble the inputs' signals more closely.

This process is repeated for all the observations in the training data set, distorting the initial arrangement of neurons. The result is a neuron arrangement that conforms to the given input data. It should be noted that the geometry of the grid is not changing, just that the grid is getting deformed. That interactive comparison and adjustment on the weights represents the competitive learning and creates the self-organizing map, creating a good representation of the clusters based on the input observations.

The self-organizing map algorithm has the following steps:

1. Randomly initialize all weights.
2. Select input vector x = [x1, x2, x3, ..., xi], where xi are the input variables.
3. Compare x with weights wj for each neuron j to determine the winner neuron.

4. Update the winner neuron so that it becomes more like x, together with the winner's neighbors.
5. Adjust the parameters: learning rate and neighborhood function.
6. Repeat from step 2 until the map has converged, which means no noticeable changes in the weights occurs or a pre-defined number of training cycles have passed.

One of the advantages of self-organizing maps is that the algorithm attempts to map weights to conform to the input data with the main objective of representing multidimensional data in an easier and more understandable form. That makes it easier to visualize and understand the distribution of the input data.

Another benefit is that a self-organizing map can be used as a technique to reduce the dimensionality of the input data. The final grid of clusters makes it easier to observe similarities in the data.

In the machine learning community, self-organizing maps are getting more popular, especially considering the continuously increasing amount of data available for analytical model development. Self-organizing maps are fully capable of clustering exceptionally large and complex data sets. On the other hand, with the current computing capacity, self-organizing maps can be trained in a short amount of time.

One of the weaknesses of self-organizing maps is that the neurons require a substantial amount of data to create meaningful clusters. If the neurons are provided too little information or too much extraneous information, the clusters might not be accurate or informative.

Another weakness assigned to self-organizing maps is the assumption that nearby data points behave similarly. If this assumption is violated, the self-organizing maps will not converge to useful and meaningful clusters. As a result, at the end of the clustering process we might find a few groups with most of the observations from the input data set and several groups with few cases.

Use Case Study: Insolvent Behavior

Insolvent behavior among customers can generate millions of dollars in losses to companies. However, not all insolvent customers have the same characteristics. To save the company money, self-organizing maps could be used to highlight good insolvent customers among all the insolvent. As mentioned before, customer behavior can consist of several dimensions, considering multiple types of attributes. Most of the clustering techniques find it difficult to handle a great amount of input variables. Self-organizing maps are capable of handling not just substantial amounts of data, but also high dimensionality in the input space and complex data. Considering customer behavior, particularly assigned to unmanageable debt, here are some of the possible attributes:

- Payment behaviors in terms of percentage of the invoice amount
- Frequency and recency of payment delays
- Customer demographics
- Amount and period in debt

- Products and services assigned to the debt
- Customer aging, segmentation, status
- Credit score

Self-organizing maps can be used to generate clusters of customers according to their insolvent behavior. The clusters can be used to understand the risk and harm of each group of customers to the company operation cash flow. From the insolvent customers' group, it is possible to identify the ones that are less harmful to the company. Some groups reveal very harmful customers. Other clusters point out customers in between those behaviors.

Good insolvent customers might present a payment behavior that causes low damage to the company while they still consume its products and services. A proper relationship with these customers can bring lots of benefits to the company. For example, there are some customers that recurrently delay their payment for a brief period and then make their payment, even the minimum amount. Once the customer becomes insolvent, the company immediately sends out a collection letter. This subset of customers pays their bills whether they receive the letter or not. The company then is just wasting money in preparing and sending out those letters. A simple classification model upon the insolvent clusters can identify the customers that will likely pay their bill regardless of the collection letter. The company can then avoid issuing the letter for those customers and save a substantial amount of money.

Figure 6.12: Self-organizing Maps for Insolvency Segmentation

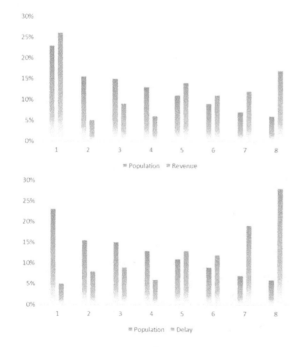

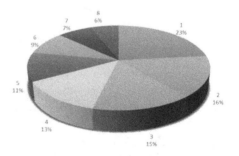

Four good groups represent 28% of the delay, 46% of the revenue, and 67% of the population.

Four bad groups represent 72% of the delay, 54% of the revenue, and 33% of the population.

Cluster Evaluation

Data scientists need creativity and collaboration when evaluating clusters. They need creativity to understand the business aspects behind the differences in each cluster and associate these main characteristics to the company's goals. They need collaboration because this process involves a lot of business knowledge, sometimes specific to a department or business action like a campaign, promotion, or relationship. Quite often data scientists need to work with business analysts from distinct departments to better understand the meaning of each cluster and then how to correlate them to business approaches.

Cluster interpretation usually involves manual inspection of the average attributes associated to each cluster and the difference to the entire population. It also might require a comparison to existing analysis previously performed. Technically, data scientists can evaluate clusters based on quality measures, or distance measures, evaluating the similarity within each cluster.

At the beginning of the chapter, we stated some pre-requirements for a good clustering processing, including meaningful attributes, numeric variables, and not too many inputs. However, when analyzing clusters, data scientists can add more variables to the analyses, variables eventually not used during the clustering, and attributes not added to the input space. These new input variables can be relevant to the business and can help in discriminating the clusters. At this point, even categorical variables that were not used during the clustering process can be used here to help identifying the main characteristics of each cluster and how they are correlated to the business.

Cluster Profiling

Cluster profiling can be defined as the derivation of a class label from a proposed cluster solution. In cluster profiling, the emphasis switches from cluster formation to cluster interpretation. The objective is to identify features or combinations of features that serve to uniquely describe each cluster of interest.

Cluster profiling attempts to assign a distinguishing class label to each of the derived clusters such that it uniquely identifies the cluster. Here is one straightforward profiling methodology:

1. Cluster the sample using a hierarchical or partitive clustering technique and save each cluster's identifying number.
2. Fit a decision tree or a regression with the cluster number as the target to identify which of the predictor variables are different from cluster to cluster. These variables, either individually or in combination, are the basis for meaningful cluster descriptions (class labels).

One strength is that cluster profiling can aid in a marketing process. Data scientists can build customer segments to understand how to best market a product or set of products to each customer group.

The cluster profiles might not be very robust. For example, if the data scientists use a different subset of the data, use different clustering methods, or use a different number of clusters, the cluster profiles could be different.

Cluster profiling can lead to marketing segmentation. These are some of the predictor variables used in the cluster analysis:

- Demographic – age, gender, income, education, and ethnicity
- Geographic – ZIP code, city, climate, urban, suburban, and rural
- Lifestyle – interests, opinions, and social class
- Behavioral – purchasing habits, brand loyalties, and price sensitivity

When the clusters have been profiled, data scientists can customize the marketing message to be more unique to a specific audience. For example, instead of sending out a generic, mass marketing message, data scientists can target the advertisement to consumers who are more likely going to buy the product. This can boost customer acquisition through advertising and keep costs lower.

Additional Topics

With customer segmentation, predictive models that are constructed on the segments often produce more accurate predictions than are achieved when a single model is constructed using the entire sample.

Cluster segments can also be used as predictor variables in a predictive model such as logistic regression. Using the cluster segments might improve the predictive accuracy of the model.

Although it is usually assumed that clustering produces interpretable results, this is not always the case. The clustering algorithms might produce uninterpretable results. The interpretation is user defined.

The decision about which variables to use for clustering is a crucial decision that will have a significant impact on the clustering solution. A poor choice of variables leads to an uninterpretable cluster profile.

Summary

This chapter covered a common analytical approach in data science – clustering. The most popular methods, hierarchical, k-means, and self-organizing maps, were introduced along with business cases as examples. Unsupervised models consist of a very particular approach in data science, especially because they are a type of model where a target is not present. It makes it hard to evaluate the model results in terms of accuracy and in terms of applicability. However, unsupervised models are quite common in data science as an analytical solution to business problems. They are easy to understand and easily translated into business rules.

The main goal of unsupervised models is to search for insights and to highlight the main characteristics in a data set, grouping similar observations into the same clusters. Clustering models are frequently one of the first analytical approaches developed by data scientists when looking at customer behavior, product and services consumption, and any type of data set segmentation. It might be used as a method to homogenize a data set before training a supervised model. For example, cluster diverse types of churn and then train one supervised model for each cluster. This approach ends up with better overall accuracy than a single model fit on the entire data set.

The next chapters will continue to explore unsupervised models, focusing on association rules, link analysis, path analysis, network analysis, and text mining.

Additional Reading

1. Collica R.S. 2017. *Customer Segmentation and Clustering Using SAS Enterprise Miner*, Third Edition. Cary, NC: SAS Institute Inc.
2. SAS Institute Inc. 2017. SAS Enterprise Miner 14.3. Cary, NC: SAS Institute Inc.

Chapter 7: Unsupervised Models–Semi Structured Data

Chapter Overview

In this chapter, we continue to learn about unsupervised models, which are models without a target. This type of model aims to better understand patterns in data and to raise insights about it. It is used to highlight customer behavior, patterns in transactions or relationships, and associations between items. It is quite often applied to structured data, like transactions, and unstructured data, like text data.

In this chapter, we will learn specifically about association rules analysis, a method assigned to market basket analysis applications. We will also learn about path analysis, a method to understand events that take place along steps or sequence of steps in a timeline, and link analysis, a method to understand how events are correlated to each other regardless of a timeline. Finally, we will learn about text analytics, a technique to analyze texts and to extract insights and knowledge from it.

The main goals of this chapter are the following:

- Explain and apply association rules analysis to create market basket analysis.
- Explain and apply path analysis to solve business problems from a timeline perspective.
- Explain and apply link analysis to solve business problems where events might be associated to each other.
- Explain and apply text mining models to solve business problems where unstructured text data is available.
- Describe an unsupervised model like an association rules model.
- Interpret the results of various unsupervised models.

Association Rules Analysis

Association rules analysis is a method to discover events, elements, or items that occur together in a data set. If timestamp information is available, the association rules analysis turns into a sequence association analysis, where the sequence of the events taking place is relevant. Association rules analysis or sequence association rules analysis aims to find interesting relationships between items in a database. These relationships are described as implication rules or items that imply other items. For example, in a point-of-sale system for a supermarket, association rules analysis might discover an implication rule like {wine, cheese => bread}. This rule implies that a customer purchases wine, cheese, and bread together, or more specifically, if a customer purchases wine and cheese, they also purchase bread. Because the association rules analysis method does not consider the time, the rules discovered are symmetric, which means that wine and cheese implies bread and bread implies wine and cheese. The data scientist needs to analyze all these rules to identify the ones that make the most sense to be used to boost sales. There are some statistical metrics that can be used to evaluate the importance of the rules, and the data scientist should use them in conjunction with business knowledge.

Market Basket Analysis

Association rules analysis is often applied in projects assigned to market basket analysis. Market basket analysis is used to analyze streams of transactional data for the combinations of items that occur more commonly than expected. Data scientists can use this as a technique to identify interesting combinations of items purchased and use these rules to boost sales. Items in the same rule are more likely to be purchased together and might be offered in conjunction during a campaign or be placed together in a supermarket.

Market basket analysis is applied in many industries for multiple business problems. Some combinations of items might be very intuitive. For example, customers who purchase bread also purchase butter. Some rules might be less intuitive and be used to boost sales. For example, customers who purchase data packages in mobile devices also consume streaming content on demand.

Sequence analysis is an extension of the association rules analysis where a time dimension is available in the transactional data. In this way, transactional data is examined for sequences of items that occur more commonly than expected. In the sequence analysis, the order of the events is quite important. The sequence of items purchased by customers in a specific time frame is an example. Data scientists can use sequence analysis to identify patterns of purchase in a timeline. Items within a sequence rule are purchased in an order and can be offered together or In a specific sequence during a sales campaign. Sequence analysis can also reveal problems in transactions occurring in a timeline. Steps performed by fraudsters in cyber-attacks or problems experienced by users when browsing a website is an example. Sequence analysis is discussed in more detail in the next section.

Confidence and Support Measures

In the simplest association rules analysis situation, the data consists of two variables: a transaction and an item. For each transaction, there is a list of items. Typically, a transaction is a single customer purchase, and the items are the things that were bought. An association rule is a statement of the form (item set A) => (item set B). In other words, an association rule is a pattern that states when A occurs, B occurs with a specific probability.

The aim of the association analysis is to determine the strength of all association rules among a set of items. The strength of the association is measured by the support and confidence of the rule. The support for the rule A =>B is the probability that the two items occur together. The support of the rule A => B is estimated by the following equation:

$$\frac{transactions\ containing\ A\ and\ B}{all\ transactions}$$

The support measure is the evidence that A and B occur together when considering all transactions within the database. It is important to notice that the support measure is symmetric. That is, the support of the rule A => B is the same as the support of the rule B => A. For example, the support of the rule bread implies butter has exactly the support of the rule butter implies bread. The support determines how frequent this rule occurs in the data set or, in other words, how relevant it is to combine these items (bread and butter) into a campaign or promotion.

The confidence of an association rule A => B is the conditional probability of a transaction containing the item B given that it contains the item A. The confidence is estimated by the following equation:

$$\frac{transactions\ containing\ A\ and\ B}{transactions\ containing\ A}$$

The confidence measure is the extent to which that A occurs with B in transactions that include A within the database. Notice that unlike support, confidence is asymmetric. The confidence of A=>B might be different from the confidence that B=>A, depending on the number of occurrences of A and B together and alone in the transactions.

Use Case: Product Bundle Example

For example, consider a telecommunications company. This company provides many items that customers can purchase and consume such as broadband, streaming content, mobile lines, data packages, and international minutes. In this example, there are five transactions to analyze. Considering the items purchased together by the different customers as seen in Figure 7.1, a few rules can be identified with their respective support and confidence:

1. Broadband implies data package, with a support of 2/5 and a confidence of 2/3. If you observe all the transactions carefully, broadband and data package occurs twice within the five transactions (second and fourth transactions). The support is 2 out of 5, or 20%, meaning that these two products are purchased together in 20% of all purchases. The data package is purchased three times (second, third, and fourth transactions). The confidence is 2 (transactions with broadband and data package) out of 3 (transactions with data package), 66%, meaning that broadband and data packages are purchased together 66% of the time when a data package is purchased.
2. Mobile line implies broadband, with a support of 2/5 and a confidence of 2/4.
3. Broadband implies mobile line, with a support of 2/5 and a confidence of 2/3. Here we can see an example of a symmetric rule: mobile line implies broadband, and broadband implies mobile line. The support as expected is the same, 2/5 for both. However, the confidence is different: 2/4 for the first rule and 2/3 for the second one. The difference is how many times broadband and mobile lines occur individually considering all the transactions.
4. Streaming content and mobile line imply data package, with a support of 1/5 and a confidence of 1/3. Here it is important to notice that the size of the rules can vary. We might find rules with multiple items on the left side of the rule (antecedent) and multiple items on the right side of the rule (consequent).

Figure 7.1: Association Rules for Product Bundle

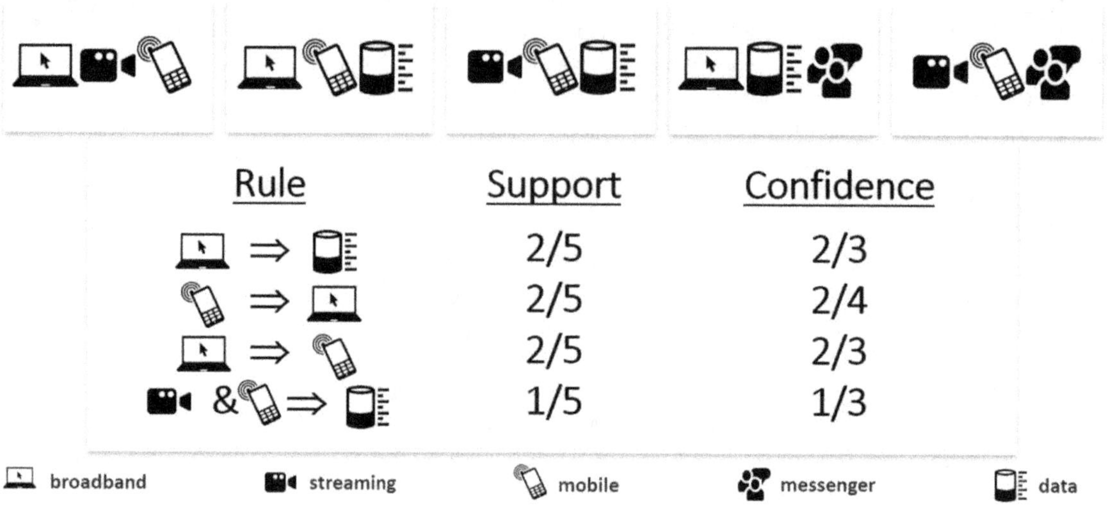

Rule	Support	Confidence
💻 ⇒ 📊	2/5	2/3
📱 ⇒ 💻	2/5	2/4
💻 ⇒ 📱	2/5	2/3
🎥 & 📱 ⇒ 📊	1/5	1/3

broadband   streaming   mobile   messenger   data

Expected Confidence and Lift Measures

Support and confidence are good measures to evaluate the strength and relevance of the association rules discovered. Support informs how frequently the items in the rules occur together out of all transactions in the database, and confidence informs how frequently some items in the rules occur together once a subset of them occurs.

Because confidence is asymmetric, which means that item A implying item B is different from item B implying item A, it is important to know whether a specific item is really causing the occurrence of another item. Some items might be extremely popular in the database, and it is difficult to evaluate if this item would occur in the transactions because of itself or because of the occurrence of the other items.

To address this issue, other measures to evaluate association rules can be used. Expected confidence and lift are two of them. The expected confidence of the rule A => B is estimated by the following equation:

$$\frac{transactions\ containing\ B}{all\ transactions}$$

The expected confidence is the confidence that would be expected if the rule A => B is false. For example, if A and B are unrelated, we can expect that the confidence of item A implies item B would be based on the relative proportion of the item B from all the transactions. If the expected confidence of item B (occurrence of item B from all transactions) is greater than the confidence of item A implies item B (occurrence of items A and B out of all transactions with item A), then we can assume that item A does not imply item B. Item B would occur many times no matter the number of times item A occurs.

The second measure to evaluate the strength and importance of the association rules is the lift. Lift is associated with both confidence and expected confidence measures. The lift of the rule A => B is the confidence of the rule divided by the expected confidence. The lift of rule A => B is calculated by the following equation:

$$\frac{\textit{confidence of A => B}}{\textit{expected confidence of A => B}}$$

The lift can be interpreted as a general measure of association between the two item sets in the association rule. In practical terms, the lift can assume values greater than 1, less than 1, or equal to 1. Values greater than 1 indicate positive correlation, which means the item A really implies the item B. Values equal to 1 indicate zero correlation, which means items A and B would occur independently. Values less than 1 indicate negative correlation, which means item A does not imply item B, or item B would occur regardless the occurrence of item A. Notice that lift is symmetric. That is, the lift of the rule A => B is the same as the lift of the rule B => A.

Association Rules Analysis Evaluation

Association rules analysis is especially useful in marketing and managing product portfolios in addition to market basket analysis applications. However, the interpretation of the implication in association rules analysis can be a bit risky. High confidence and support do not exactly imply cause and effect. Customers do not purchase butter just because they purchased bread. They purchase both butter and bread together. Furthermore, the rule might not be necessarily interesting. For example, customers acquiring debit cards are also acquiring credit cards. In this scenario, it is a prerequisite for the financial institution that customers need to have the debit card first to have the credit card. That would be a strong rule with high value support and confidence, but it would be useless since it is a business rule, not an insightful association rule. Furthermore, the two items occurring together out of all the transactions might not even be correlated. Data scientists need to always work together with the business departments to interpret the results of unsupervised models like association rules analysis.

Consider another example of the association rule Mobile => Broadband in Figure 7.2. This rule has 50% support (5,000 mobile and broadband purchases together / 10,000 transactions). This rule has 83% confidence (5,000 mobile and broadband purchases together / 6,000 mobiles). Based on these two measures, this might be considered a strong and relevant rule. Half of the transactions have this set of items, and the confidence of purchasing a broadband after purchasing a mobile line is about 83% of the time. If the two products are independent, then knowing that a customer has purchased a mobile line does not help in knowing whether that customer has purchased broadband or not.

The expected confidence if the two accounts were independent is 85% (8,500 broadband purchases / 10,000 transactions). The lift is 0.83/0.85 or 0.98. Because the lift is less than 1, the association rule is not useful. It means that customers would purchase broadband anyway, no matter if they bought the mobile line or not. Furthermore, those without a mobile line are even more likely to have broadband (3500 broadband purchases without mobile / 4000 mobiles), with confidence of 87.5%. Based on that, mobile and broadband are negatively correlated as shown in Figure 7.2.

Figure 7.2: Association Rules Evaluation

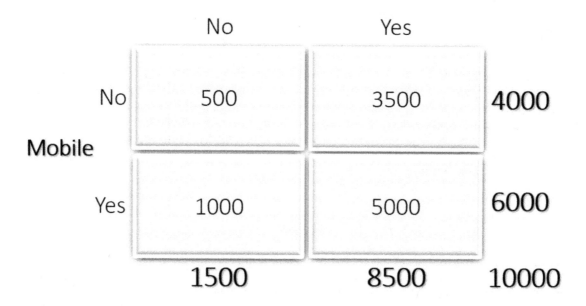

Support (MOB ⇒ BRO) = (5000/10000) **50%**

Confidence (MOB ⇒ BRO) = (5000/6000) **83%**

Expected Confidence (MOB ⇒ BRO) = (8500/10000) **85%**

Lift (MOB ⇒ BRO) = 0.83/0.85 < 1

A particularly critical point in any data analysis is that the data is not generated to meet the objectives of the analysis. The data is generated to meet the needs of the business goals. Data scientists must determine whether the data, as it exists, has the capacity to meet the objectives of the analysis and then accomplish the business goals. Quantifying relationships among items occurring together would be useless if very few transactions involve multiple items. If a database does not have enough data about relationships among items, for example, by not having enough baskets with not enough items within the baskets, data scientists will not be able to find any meaningful rules to see when items are purchased together or to see when customers purchasing some items imply purchasing others. Therefore, it is important that data scientists do some initial examination of the data before attempting to do market basket analysis. In general, this initial examination is required before attempting any analytical model, either supervised or unsupervised. It is important to know that the data required to create models are available, meet the business goals, and are enough to model the problem.

One strength of association rules analysis is that it is easy to implement and interpret the results. Association rules can also search through exceptionally large data sets of random data to identify diverse types of patterns. It is a technique ideal for categorical data.

One weakness is that many items can lead to exceptionally large rule sets that are difficult to comprehend and manage. Some filtering might be required to generate a reasonable number of rules. Ranking the rules based on the measures can be useful as well, for example, when selecting the top rules based on lift and support.

The problem is when a minimum level of support is used to filter the association rules. The assumption is that all the items have similar frequencies or are of the same nature, or both. In many situations, some items appear frequently in the data and others rarely appear. If the frequencies of the items vary to a large degree, then rare items might be left out of the association rules if the minimum support level is set too high. Sometimes the rare items are the most valuable items in the database. On the other hand, if the minimum support level is set too low, the number of association rules discovered might become unmanageable.

Finally, if there are many items, it is difficult to determine the right number of items. Supermarkets and grocery stores might have thousands of items. The number of association rules might be exceptionally large, and the rules might be extremely complex. Filters in the measures of the rules and data categorization might be useful in searching for affinity patterns among the items.

Use Case: Product Acquisition

Market basket analysis can be used to identify what products and services are purchased together. After these association rules are established, data scientists can recommend products and services to customers based on what products are usually purchased together.

For example, a telecommunications provider can offer a set of products and services, such as streaming content, data packages, fixed and mobile lines, broadband, fiber, and on-demand services like security and tech support. This provider can search associations among products and services through customer purchase history information: what phone and internet services they purchased, what streaming and digital content they used, what contract they have, what method of payment they used, and what on-demand services they consume, among others.

The resulting association rules can identify opportunities in cross-selling and up-selling campaigns. Quite often, companies encourage their customers to spend or consume more by recommending related products that complement what is being bought already. These recommendations can be based on the association rules. For example, customers who purchase streaming movie content, plus have month-to-month contracts, have multiple lines, and purchase fiber. Based on this association rule, this provider can offer fiber to all customers who have that set of products, streaming movies, month-to-month, and multiple lines. They are more likely to purchase fiber than the other customers who have not purchased these products already. Another association rule could be customers who purchased broadband, have a two-year contract, bought tech support, and online security on-demand services. Finally, a third association rule could be customers who have purchased online backup also purchased tech support and streaming content.

As you can notice, the size of the association rule can vary on both sides, on the left side (antecedent) and on the right side (consequent). The left and right sides of the rules are used to identify patterns when customers purchase and consume products and services. The left side of the rule is used to identify the customers who will be added to the cross-sell and up-sell campaigns.

Sequence Analysis

Sequence analysis or sequence association rules analysis is an extension of the association rules analysis or market basket analysis. This extension in the technique includes a time dimension to the analysis. The addition of time restricts the possible associations because it is now based on the time sequence. In this way, historical transaction data are examined for sequences of items that occur more commonly than expected. Data scientists might use sequence analysis to identify patterns in purchasing or consuming, or even to detect problems in events based on time.

An easy comparison between these two techniques (association analysis and sequence analysis) and the data required to perform those different analyses is supermarkets. Imagine you are shopping at a grocery store. You walk throughout the aisles and grab the products that you want and put them in the shopping cart. Once you finish, you go to the cashier. You need to remove the products from the shopping cart and arrange them before starting the scanning. Perhaps as you organize them you set them based on categories, or weight, or refrigeration need, and so on. You finish your check-out and leave. The grocery store now knows what you purchased and can match that to what other customers bought. But this grocery store does not know the order you selected your products. It does not know what you purchased first, what was more important, and what you purchased last, as consequence of previous products. However, knowing what customers purchase together is already a big deal. This grocery store can run an extremely useful analysis based on association rules to better understand what products are purchased together and boost sales by displaying products together or defining combined sets of products with discounts.

Now suppose that this grocery store also has a website for online shopping. It is able to know the exact order the products were selected and put in the virtual shopping cart. The timestamp of selecting the products on the website distinguishes the order of selection. That makes an enormous difference in some rules. For example, based on the data on the physical store, one association rule can be {wine and cheese} implies {bread and crackers}. However, in association rules, these rules are symmetric, so {bread and crackers} implies {wine and cheese}. The sequence association rule or sequence analysis based on the data from the virtual store can define exactly the implication on the rule. If customers often select first wine and cheese and then they select bread and crackers, the sequence analysis raises the rule {wine and cheese} implies {bread and crackers} and not the other way around.

The most important thing in sequence analysis is that the data consists of customer transactions over time. Each transaction is an item purchased at a certain time associated with a customer. The goal is to extract rules in a proper sequence. For example, sequence analysis might show that 60% of customers who first buy item A also purchase item B within three weeks. This information can be used in cross-sell and up-sell campaigns. Not just the information about the association, item A implies item B, but also the information about the time frame, item B is purchased on average three weeks after item A is purchased. That information can show not just the customers to be targeted and the products to be offered, but when the campaign should be launched.

Sequence analysis has been used extensively in web usage mining for finding navigational patterns of users on the website. It has also been applied to finding linguistic patterns for opinion mining.

As observed for the market basket analysis, sequence analysis is easy to implement and interpret, and it is ideal for categorical data.

One of the weaknesses of sequence analysis is also the same as market basket analysis. If there are many items to be analyzed, the number of rules can be huge and lead to useless rules. Similarly, rare items might be left out if the minimum level of support is set too high. It is also difficult to determine the right number of items if the number of available items is exceptionally large.

Use Case: Next Best Offer

Sequence analysis can be used by data scientists to identify what products and services are purchased together and in what specific sequence in time. After these sequence association rules are established, data scientists can use the results to decide what is the next best offer for each customer considering the previous purchases.

For example, a telecommunications provider can offer a set of products and services, such as streaming content, data packages, fixed and mobile lines, broadband, fiber, and on-demand services like security and tech support. This provider can also have a sequence variable in the transaction data that identifies when the product or service was purchased. This data can then be used to search for associations among products and services through customer purchase history information in a timeline.

The results of the sequence analysis can identify opportunities for the next best offer campaigns. The recommendations on the next best offer are based on the sequence rules. For example, customers who purchased fiber optic service also purchased a streaming movies package on average four weeks later. Based on this sequence rule, this provider can offer streaming movies to all customers who have purchased fiber optic service in a range of four weeks. These customers are more likely to purchase a streaming movie package than the other customers who have not purchased fiber optic service. Another sequence rule could be that customers who purchased multiple lines also purchased streaming television on average six weeks later. Finally, a third sequence rule could be that customers who have purchased device protection also purchased online backup on average two weeks later.

Analogous to the market basket analysis, the size of the sequence rule can vary on both sides – on the left side (antecedent) and on the right side (consequent). The left and right sides of the rules are used to identify the patterns when customers purchase and consume products and services in a timeline perspective. The left side of the rule is used to identify the customers who will be added to the next best offer campaign. The right side of the rule is used to identify what products to offer to each customer.

Link Analysis

Link analysis is the process of discovering and examining the connections between items in a database. It considers both tabular and graphical representations of the associations, sequences, and networks that exist within a database. Link analysis aims to discover patterns of activities that can be used to derive useful conclusions about different applications in many industries such as fraud detection, computer networks, traffic patterns, and web usage.

Link analysis uses different techniques and algorithms to identify, explore, and visualize connections between entities. These entities might be people, cell phones, vehicles, organizations, countries, transactions, or machines. The main goal of link analysis is to reveal patterns of relationships among these entities. For example, many retailers now capture every bit of a transaction and interaction that customers have with them. Discovering and understanding these relationships and connections within this big data can provide valuable insights to help businesses make better decisions.

Link analysis can reveal associations between items that are frequently purchased together. In some situations, we want to know what items have infrequent or no associations. The lack of relationship is the association that we are looking for. Imagine a strong association between products A and B and between products C and B. These strong relationships between these products reveal that product A is frequently purchased together with product B, and products C and B are also frequently purchased together. However, there is no association between the products A and C. Suppose that product A is Coke®, product C is Pepsi®, and product B is potato chips. Customers who purchase a cola product also buy potato chips. However, they do not purchase distinct brands of cola products at the same time. These products A (Coke) and C (Pepsi) are surrogate products. A vendor can use this information to bargain for a better sales price for either product from the provider. This example is very intuitive by knowing cola products and potato chips. However, think about supermarket chains having thousands of items in their databases. Finding surrogate products is not a trivial task, and link analysis can help them with this challenge.

The results of a link analysis are a graph network that visualizes connections between distinct types of information. Besides retail, the following are many other applications of link analysis in different industries:

- Police departments can use link analysis for criminal investigations and crime detection.
- Financial institutions can use link analysis to investigate hidden connections between fraudulent customers to build a fraudsters profile.
- Federal government agencies can use link analysis to track connections of terrorists and other suspicious people in social media.

Figure 7.3: Links Connecting Entities

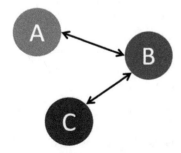

- Telecommunications companies can use link analysis to understand how customers relate to each other and how they influence other customers in some business events like churn and product adoption.
- Insurance companies can use link analysis to establish relationships between claims and to identify suspicious connections between entities on these claims.

One of the benefits of link analysis is to produce visualizations as outcomes. These visualization outputs are often very intuitive and extremely easy to understand. Data scientists can easily identify customer behavior associated with business events and discover fraudulent activity within large transaction data sets.

Figure 7.4 shows a network represented by nodes (circles) and links (connections between circles). Nodes can describe cell phones and links can describe calls. Varied sizes for the nodes can represent ARPU (Average Revenue Per User). Different shades of color for the nodes can represent distinct communities within the network (the way cell phones are grouped together based on the frequency and duration of their calls to each other).

Even though the link analysis can be used in large data sets, its outcomes can be better visualized when the network is not too large. It is common to filter out uninteresting transactions and relationships from the output so that data scientists can focus on associations with a specific business purpose. In this way, the outcome from link analysis can work best in situations where there is a limited number of transactions or relationships. Its visualizations begin to deteriorate with many transactions because the graph becomes too populated and then the visual analysis becomes harder to interpret.

Figure 7.4: Link Analysis

Use Case: Product Relationships

Link analysis is widely used in the retail and communications industries. The essential variables are the customer identification number and the items purchased. Unlike association rules analysis, the transactions in link analysis do not need to be in the same data set or under the same transactional identification. The transactions do not need to even be in the same time frame. Any relationship between customers and items purchased can be used to create the network to describe the relationships between customers and products.

Data scientists can choose to see which customers have bought a specific product and service or what products and services a specific customer has bought. For example, consider content in streaming television services. A network of movies and customers are easily created based on the transactions of what movies were watched by which customers. Another important network would be the relationships of movies based on the common customers. For example, customers A, B, and C watched movies 1 and 2. There is then a relationship between movies 1 and 2 with a strength of 3 (3 customers watched them). Customers D and E watched movies 3 and 4. There is then a relationship between movies 3 and 4 with strength of 2. An exceptionally large network can be created considering a big data set of movies and customers. However, even in a large network, it would be easier to highlight strong relationships between movies watched by common customers and then use these strong relationships to recommend new movies to customers.

Path Analysis

Path analysis evaluates a chain of consecutive events that a given user or entity performs during a specific time frame. This is not the same statistical "path analysis" that Sewall Wright developed in 1921 that is a precursor to structural equation modeling. It is a way to understand users' behavior to gain actionable insights into the data that accounts for temporal events. For example, in the online retail industry, path analysis can be a key data science technique to understand the consumer's behaviors when purchasing over time. If a retailer sells products online, that company wants to convert as many visitors to a completed purchase as possible. Data scientists can use path analysis to determine what features of the website encourage customers to navigate until the check-out. They can also look for big drops in the navigation, associated to paths or features on the website that lose or confuse potential customers.

Data scientists can use path analysis to examine the customer experience with online games. For example, if a large amount of time is spent around the menu page, there might be a problem with the user experience or the structure of the website in terms of the information provided. By following the customer path, data scientists can detect problems in the website flow and adjust the website structure as needed.

Data scientists can also use path analysis to customize an e-commerce shopping experience. Analogous to the link analysis, the temporal relationships between customers and products are identified by the path analysis and can be used to recommend products and services to customers based on the similarity of the relationships found. Customers purchased products over time, and these products are related based on the customers. Customers who purchased some specific products can receive recommendations to buy strongly related products in the future.

There are many path analysis applications in fraud detection and cyber security. The sequence of steps performed by the fraudsters before committing the fraud is crucial to understand the modus operandi in fraud events. The frequency and recency of some of the steps in that sequence can be used to alert breaches in the customer's behavior when consuming products or services and making some transactions.

Finally, path analysis can help solve performance issues on a server. Data scientists can examine a path and realize that its site freezes up after a certain string of events. The error can then be documented and fixed.

As in link analysis, one of the greatest strengths of path analysis is the output. The graphs produced as the results of the path analysis are extremely easy to interpret and understand. The graph is a visual portrayal of every event performed during a time frame and can be used to analyze, investigate, and understand customer behavior or the pattern in sequences of events. Data scientists can identify problems in the sequence of events by analyzing the path and the progression of events leading to the problem.

Path analysis describes what customers consume over time. For example, in Figure 7.5, most of customers first purchase a device, then they acquire an internet fiber connection, then multiple network devices (to create a mesh network at home for example), then online content, then streaming packages, and so on. This analysis allows companies to anticipate customers' needs and boost cascade sales.

However, just as in link analysis, one of the problems with the path analysis outcomes is that the diagrams cannot handle many activities. If the data scientists are using web log data for example, the data sets could be huge because of the massive amounts of traffic to a website and the amount of

Figure 7.5: Path Analysis

information stored in each record. Records with missing Internet Protocol (IP) addresses and changes to website content cause messy data. Finally, path analysis does not explain why customers do what they do. In some analyses, data scientists might not be interested in why some events happen, but when they happen. However, in other analyses, why the event occurred could be important, in addition to when the event occurred or what event occurred. For example, data scientists might see a big drop-off from the product web page to the purchase product page, but more work needs to be done to find out why the customers are not purchasing the product.

For both link and path analysis, when many transactions are involved, an extensive data preparation process might be required. For example, when data scientists are investigating money laundering, the number of entities can be huge as well as the number of transactions. The time frame is usually extensive too, implicating that even more data needs to be analyzed.

Finally, again for both link analysis and path analysis, when rare events are the target, these techniques might not be the best ones. Very often these techniques highlight cases and transactions based on frequency and recency, and rare events might be missed in the analysis.

Use Case Study: Online Experience

Path analysis can be used to investigate and understand customer behavior on a retail website. Data scientists can examine the customers surfing on the website, the sequence of the visits, the time spent on the web pages, the links clicked, and so on. Data scientists can then evaluate when customers are searching for specific products and services, when they are checking for product availability, and when they are comparing plans, offerings, and prices. The goal of the analysis is to detect bottlenecks, drop-offs, and effective conversions when customers reach the check-out and purchase the products. If the data scientists see a significant decrease from the product information pages to the check-out purchase page, it could be indicative of some problems with the product information page, the price, availability, quality, or even the possible methods of payment available. Data scientists then can analyze these results with the business department and figure out solutions to address the drop in customers navigating from the product information page to the check-out.

Text Analytics

Text is a type of unstructured data that can be extremely useful in data science applications. Data scientists use text analytics to analyze text data, extract relevant information, and turn raw data into business insights or even into new variables to be used in supervised or unsupervised models. Text analytics can be used to create a set of predictor variables that contain information about what is written in the text. It can also be used to cluster the topics that are mentioned in the documents.

One of the main advantages of text analytics is the ability to make sense of unstructured information. With the growth of unstructured data, particularly digital data, the importance of text analytics continuously increases. By applying text analytics on a set of documents, data scientists search and evaluate terms contained in these documents and create new features that can be used in other analytical models, like clustering or classification. Text analytics is a key field in data science with many techniques and tools, all designed to extract information from raw text data.

Figure 7.6: Word Cloud

Unstructured and semi-structured text content within organizations is growing at an unprecedented rate from applications, records, and business processes to external documents, web pages, blogs, forums, and so on. Unstructured text refers to plain text documents. Semi-structured text refers to text that has been augmented with tags, such as the text in web pages formatted as HTML files. These files have special tags to format the content in a singular way. If documents are preprocessed with tagging to include specific information, this might facilitate efficient information retrieval. XML (extensible Markup Language) and JSON (JavaScript Object Notation) represent two common languages that accommodate the encoding of attributes within a document. Many organizations store and manage this information based on data type by manually source tagging the content.

The amount of unstructured text data is massive, particularly in the digital world. To process all this information, a sequence of tasks with specific goals should be put in place. Each step handles a specific technique to prepare the data to be analyzed, to select the parts of the text that will be analyzed, and to analyze the text. For example, text parsing processes the textual data into a term-by-document frequency matrix. This step is very computationally intensive. The text filter removes all parts of speech that do not contribute to the identification of anything semantic in the document. For example, all articles such as a, an, and the can be removed from documents prior to analysis with a filter. Text topics and text clusters aim to cluster the documents into groups based on similarity. The major difference between topics and clusters is the relationship between the documents and the groups. One document can belong to multiple topics, for example, sports and journalism. One topic comprises multiple documents. However, one document can belong to one single cluster (either sports or journalism). One cluster comprises multiple documents also. The coefficients created during the topic and cluster identification can be used to classify new documents. These new documents are classified based on the similarity to the documents already fit into topics and clusters.

Text analytics is a vigorous field of research with many applications. Text analytics outcomes can be added to traditional analytical modeling to better solve a business problem. Text analytics outcomes often include relevant textual data to help data scientists find solutions for business problems, whether shedding light to a problem or producing new variables to better explain the problem.

Natural Language Processing (NLP) is a concept intrinsically correlated to text analytics. NLP can be used as a solution in some applications like Siri or Alexa or as a technique within text analytics software. In text analytics software, most of the tasks used to produce a text analytics solution rely on NLP. The primary role of NLP is to efficiently parse the document collection. Parsing creates the term table that is used when deriving topics or clusters. It is also used in sentiment analysis to understand positive, negative, and neutral levels of interactions. NLP is used to identify multiple categories in the document collection. Each row of the term table includes the term itself, the role of the term (the part of the speech or concept associated to the term), statistics related to the term (the frequency of occurrence of this term in the document collection), and stemming information (for example, run=running=runs=ran).

One of the biggest challenges with NLP is word sense disambiguation. Textbooks often use the word "train" to illustrate NLP challenges. "Train" and its variants can be a noun, verb, or adjective. When "train" is used in the context of a locomotive, it is a noun (the Amtrak train from San Diego arrived on time). When "train" is used relative to education, it can be a noun (data science training), it can be a verb (train the student), or it can be an adjective (a training exercise). When an NLP algorithm encounters the word "train," it must resolve any ambiguities based on the proper part of speech. The relevance of the word "train" in the text analysis can be evaluated on NLP tools like the term maps. The term table tells you that train appears in the document collection as a noun. The term map shows related terms, which helps you decide how train is used. For example, "train" might be connected to both travel and education. If the information gain for train-travel is higher than the information gain for train-education, then the document collection has more information about passengers traveling by train than students enrolling in a course.

Use Case Study: Call Center Categorization

As stated previously, text analytics can solve a problem, or it can add insights and more variables to better describe a business problem that will be solved by a traditional model. A common business problem that can be addressed by text analytics in conjunction with another analytical method is call center optimization. A call center needs to optimize the number of agents based on the types of calls they receive. Some of the calls might be extremely specific, and the agents need to be trained accordingly. This is a very traditional problem in optimization. Based on the number of calls by type received, an optimal number of agents can be found to address all those calls to reduce the waiting time for the customers. However, the call center does not know how many calls are received by type. The call center company knows the number of calls, but the agents categorize the call at the end of the service using whatever the first option is on the form that they complete after each call. It happens because the agents want to be available as quickly as possible for the next call so that they do not waste time at the end of the call.

The text analytics techniques can be used in the solution to the call centers. The customers contact the call center for a variety of reasons dealing with different topics. The call center agent requires subject matter knowledge to address these different topics. Data scientists can use text analytics to cluster the topic areas and estimate the optimal number of call center representatives for each topic. Some of the calls can be recorded for quality purposes. A specific software can transcribe these recorded calls into text. These transcripts of calls form the document collection for the text analytics project. Based on the table terms and the frequency of the terms happening together, some topics or clusters can be identified. For example, clusters can be associated with billing, cancellation, troubleshooting,

Figure 7.7: Text Analytics to Find Topics

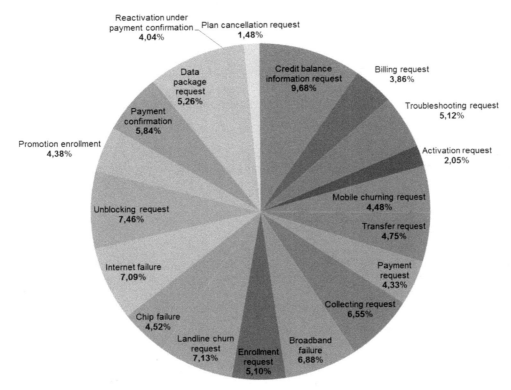

information request, or upgrades and downgrades, among others. The cluster coefficients can be applied to the entire document collection to classify each call into one of these clusters. Once this classification is done, the call center company has a much better estimate about the number of calls it receives per type and per time (hour or day). Based on this information, it can find the optimal number of agents that it needs to have available at each hour to properly address the distinct types of calls.

Before optimizing the number of call center representatives based on the subject of customers' calls, the company needs to understand the reason customers reach out to the call center. Text analysis on the transcripts of the recorded dialogs between customers and call center representatives can identify different topics. These topics can be clustered into unique groups as shown in Figure 7.7. A frequent analysis can be performed based on the groups of topics to define the best number of call center representatives by the type of information they need to be able to address.

Summary

This chapter introduced important methods in unsupervised modeling. Association rules analysis, sequence analysis, text analytics, link analysis, and path analysis have tremendous value for business applications, helping companies in any type of industry to better understand customers, processes, and relationships. What products are purchased together? What products are purchased in a specific sequence in time? What are the main complaints of the company's customers on social media? What products and services are associated together? What are the customers' behaviors when surfing on the internet? When do they drop off a web page? All these questions can be answered using unsupervised techniques.

Considering that we have much more unstructured data out there than structured data, these specific analytical approaches to handle unstructured information can be extremely valuable to most companies.

In the next chapter, we will cover two techniques in a field called network science. Network analysis and network optimization can be applied to almost any business problem, particularly considering that everything is connected somehow. Any problem can be examined in a network perspective.

Additional Reading

1. Hall, S., Osei, A., McKinley, J. 2018. "Using Market Basket Analysis in SAS Enterprise Miner to Make Student Course Enrollment Recommendations". *Proceedings of the 2018 Annual SAS Users Group International Conference*. Cary, NC: SAS Institute Inc.
2. Beasley, M. 2013. *Practical Web Analytics for User Experience*. Burlington, MA: Morgan Kaufmann.

Chapter 8: Advanced Topics in Unsupervised Models

Chapter Overview

In this chapter, you will learn about network analysis and network optimization. Unsupervised modeling started in Chapter 6 when we covered different clustering techniques. We saw the importance of clustering to segment customers into separate groups according to some similarity measure, for example, their consumption behavior. In Chapter 7, we covered a few techniques to search for patterns in transactions using association rules and sequence association rules to identify items purchased together or items purchased in a specific sequence. Also, in Chapter 7, we covered an extremely popular data science technique, text analytics. We saw how text analytics turns unstructured data into insights to solve business problems or new features to feed machine learning models. We also showed how text analytics is important considering the digital world, where increased text data is available.

Now in Chapter 8, we finalize the topics about unsupervised models with some very key techniques to understanding relationships between entities. Relationships here mean any type of relationship between entities, transactions, calls, emails, text messages, referrals, and payments, among many others. Entities can be any subject of action, a person, a cell phone, a policy owner, a bank account, a computer, and so on. The algorithms assigned to network analysis and network optimization help us find the patterns of these relationships.

The main goals of this chapter are the following:

- Understand the concepts associated to network analysis and when to apply network analysis to solve business problems.
- Understand the concepts associated with network optimization and when to apply network optimization to solve business problems.
- Describe a statistical unsupervised model using one of the network analysis or network optimization techniques described.
- Interpret the results of a network analysis or network optimization model in terms of business goals.

Network Analysis

Network analysis is the mapping and measuring of relationships and flows between nodes. The nodes in the network can be people, groups of people, companies or businesses, documents, transactions, computers, and so on. The links can be any type of relationship or flow between the nodes. Network analysis enables us to identify and understand the importance of certain nodes within a network, or the importance of certain links within the network, as well as to identify sub-networks or clusters within the entire network.

Companies use network analysis to address relevant business events such as churn, fraud, and facility location in addition to strategies related to increasing sales, product adoption, and service consumption.

Governments use network analysis to tackle terrorism, money laundering, and fraud in tax payment or in health care, among many other applications.

A common example of network analysis is social network analysis, which is the process of investigating social structures. Examples of social structures commonly visualized through social network analysis are social media networks, information circulation, friendship networks, and disease transmission. In this type of network, a person can be a node and the link can be a connection between them (a like, a tag, a message, etc.).

Network analysis allows organizations to better understand the relationship between any type of entities. For example, Figure 8.1 shows how employees communicate internally by email. Based on the frequency and recency of their emails to each other, employees can be grouped into distinct communities, represented here by distinct color shades.

Networks can be directed or undirected. In a directed network, each link has a direction that defines how the information flows over that link to connect the nodes. In an undirected network, each link has no direction, and the flow can be in either direction. For example, communication messages are a directed network where customer A calls customer B. The direction of the link describes the connection and therefore the type of the relationship. The authorship network is an undirected network. You write a paper with two or three other colleagues and then you are all co-authors. There is no direction on this type of relation. As you write more papers with someone, your relationship gets stronger as you might share more connections.

Networks can also be weighted or unweighted. Common link weights in the transportation industry are driving distance (or driving time) from one node to another node. In communications, the type and frequency of the connection can constitute the weight. For example, subscriber A calls subscriber B 20 times and sends 30 text messages and 5 multimedia messages. The sum of these connections can define the link weight between subscribers A and B. If calls, text messages, and multimedia messages have different importance values for the company, the weighted sum (based on importance) of these connections can define the link weight. Nodes can also have weights to describe their importance within the network. For example, consider paper references in a co-authorship network, an author might be especially important by the number of papers written and/or by the number of references. If as an author, your paper is referred by a low weighted author

Figure 8.1: A Social Network

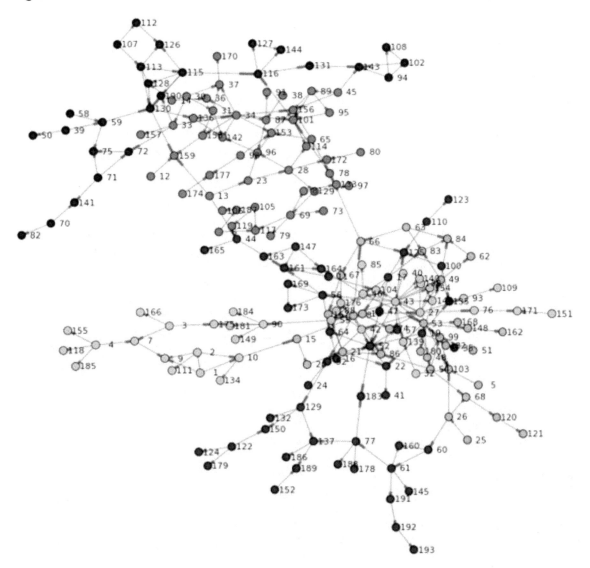

(few papers and no references), your importance does not grow that much. However, if your paper is referred by an authority in the field with high weight (lots of papers and papers references), your importance increases substantially.

In Figure 8.2, nodes can represent employees (circles), and links can represent connections (arrows). The connections can be done through emails. The relationship between employees can be weighted, according to the number of times they communicate to each other. The size of the arrows represents the frequency of emails.

Figure 8.2: Types of Networks

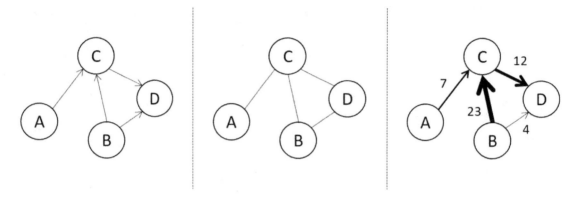

Traditional methods of data analysis usually consider individual attributes from all the observations. However, network analysis considers both the individual attributes and the relationships among the individuals. In fact, the relationships among the individuals within the network are usually more important than the individual attributes. Understanding the relationships in the network can enable companies to predict new business events and focus on customer relationship management.

Data scientists can extract mathematical properties from networks like the average distances between the nodes or the strength of the relationships or the subgroups within the whole network. All this information can help data scientists characterize complex networks and extract useful knowledge from it to solve business problems.

Although the use of graphs to describe the characteristics and structure of the network and their components such as nodes and links is straightforward and intuitive, larger networks with many nodes and links become more difficult to visualize. The network algorithms become even more important in this type of scenario, but the outcomes and the visual analysis can be more difficult to interpret. For example, a graph illustrating the network in a telecommunications company can have millions of nodes and hundreds of millions of links. Visually interpreting this network would be a challenging task. However, the network algorithms can easily find communities within that network and based on network measures identify the leaders in each community. These results can be used in product adoption campaigns, targeting the leaders to spread the message throughout their communities.

The amount of data used to describe a network can be huge. To understand the network and its structure, or the pattern of the relationships, or even the members' behavior, effective data preparation is required. Very often, during the data preparation phase, the data scientist communicates with the business departments the type of network that will be built, the elements that will be used to describe the nodes and links, and the methods to weight nodes and links based on the business information available. This information, particularly about the weight nodes and links, is essential to compute the network metrics and derive subnetworks from the main network.

Once the data preparation is finished and the network is built based on the information available for nodes and links, a set of tasks can be performed to understand the network and describe its characteristics. In network analysis, two main groups of tasks can be performed, the subgraph analysis, where diverse types of subnetworks are detected, and the network metrics calculation, where a set of measures is computed to describe nodes and links within the network or the subnetworks identified.

Network Subgraphs

From the subgraph's detection, the most common tasks are community detection, k-core decomposition, reach networks, connected components, and biconnected components.

Community Detection

Community detection recursively identifies subgraphs (*community* is defined as a unique group of nodes) within the network based on the strength of the relationships between the nodes. The main rule is that the average link weight inside the communities should be greater than the average link weight between the communities. The simple interpretation of this rule is that the community detection algorithm aims to get together nodes that are more related to each within communities.

K-core Decomposition

Core decomposition can be an alternative for community detection in some network scenarios. Core decomposition searches for groups of nodes with the same interconnectivity. The method is nested. The most cohesive group, let us say core 5, comprises all nodes with at least five connections. The second most cohesive core, let us say core 4, comprises all nodes with at least four connections. Core 4 then also comprises all nodes in core 5. Cores 3, 2, and 1 follow the same process. The idea of the nodes in the cohesive cores is that these nodes can establish a consistent flow of information throughout the network.

Reach Networks

An especially useful subgraph is the reach or ego subnetwork. The reach subnetwork is the network considering all nodes connected from and to a subset of nodes. For example, starting from a small subset of nodes – let us say the most influential nodes, which we can call leaders – the reach network algorithm searches for all nodes connected to the leaders in a single and directed connection and creates that ego subnetwork. The primary business interpretation for the reach subnetwork is a method to find all customers that can be reached by an initial lead. Starting from 1,000 leaders, how many customers can they reach out to and spread the message for the company?

Connected Components

Connected components are also extremely useful to find groups of nodes that are all connected to each other, even if a multi-step connection is required to reach someone else in the group. In a connected component, all nodes can reach out to each other in a directed step or connection or throughout several steps going through other nodes.

Biconnected Components

The biconnected component has a quite simple concept. It is a connected component that can be split into two or more connected components if a node is removed from the subnetwork. For example, imagine a group of people that are connected to each other no matter the number of steps they need to go through to reach someone else in the group. There is a person in this group that if they are removed, the group splits into two separate groups. This person could be an account executive for a company trying to sell a product to another company. This person connects sales and presales people from the vendor to business and technical people from the buyer. If this account executive leaves the vendor, this major group splits into two groups: people from the vendor and people from the buyer. A relevant concept assigned to the biconnected components is this person that can split the one group into two or more separate groups. This node is called the articulation point or bridge. Its role in the network is usually particularly important.

Network Metrics

The second group of tasks is assigned to compute the network metrics. Network metrics are used to describe the network topology and its main characteristics. It is used to understand the role of each node in the network and their importance. It is also used to understand the characteristics of the links and how they flow the information throughout the network. There are many network metrics to be computed. These metrics are called centrality metrics, as they measure how central a node is to the flow of information in the network.

Centrality measures quantify the importance of the nodes and the links in a network. Each centrality measure has a different interpretation of business importance. Some of them can be more relevant to certain industries and targeted applications. These metrics are used to rank individual nodes and links within the network. Each one of these network metrics highlights some useful information to describe the network structure and topology, the members of the network, or the connections these members have. Network metrics can be combined to create and provide different perspectives of the importance of members and connections.

Degree

Degree centrality counts the number of connections a node has. It does not consider the strength of the connection or the importance of the other nodes. Just the number of connections. In directed networks, this centrality measure also considers the direction of the connections, counting the number of incoming connections as the degree-in and the number of outgoing connections as the degree-out. The degree is the sum of degree-in and degree-out.

Influence

First-order influence centrality describes how many other nodes are directly connected to a specific node. This is commonly understood as how many friends, or connections, a node has and indicates the potential for that node to exert direct influence over the other nodes within the network. Until this point, it is like the degree centrality, just counting the number of connections. However, if the nodes in the network are weighted, the first-order influence centrality considers the weight of each node and the weight of each link adjacent to the node the centrality is computed. Now it is not just how many connections you have but also the strength of the connections and the importance of who you relate to. Sometimes the total number of connections is less relevant than the importance of the connections. The first-order influence centrality considers the relevance of the links and the nodes. For example, in a communications network, a taxi driver might have lots of connections, but all of them are very weak. An account executive might have fewer connections, but they are strong ones and most likely to especially important people. The taxi driver would have a degree centrality higher than the account executive, and the account executive would have an influence centrality higher than the taxi driver. First-order influence centrality can be used to understand the impact of a business event that requires strong relationships between the members, such as churn or product adoption.

Second-order influence centrality represents the indirect influence of the node. It measures how many friends a node's friends have. It is the friends of the friends. It helps to illustrate how many nodes can be indirectly influenced by that node. Second-order influence centrality considers the weights of the nodes and the weights of the links adjacent to a node, as well as the weight of the links to the adjacent nodes connected to that node. In other words, the importance of my friends and the strength of my connections to my friends and the strength of the connections from my friends to their friends. For example, in a cell phone network, a subscriber might not have important friends and connections, but their friends might have important friends with good connections. Their first-order influence centrality might be low, but certainly the second-order influence centrality will be high. The second-order influence centrality captures this dynamic. In terms of business applications, a node with a high value of second-order influence might indicate that the individual is a suitable candidate to spread an important message for the company.

Betweenness

Betweenness centrality accounts for how many times a node takes the shortest path within the network. The shortest path is the shortest way to get to one point in the network from another. "Shortest" might have different connotations such as distance, importance, cost, and so on. In optimization, we usually try to minimize the cost, the time, or the distance. In this way, the shortest path from node A to node E would be the one costing less, or taking less time, or traveling the shortest distance.

Suppose that to reach to **E** from node **A**, a message needs to pass through other nodes. Let's assume the possible paths are **A→B→E** and **A→C→D→E**. This concept is especially important in optimization. For the first path, suppose that the distance between **A** and **B** is 10 units, and the cost to take this path is 25 units. The distance between **B** and **E** is 5, and the cost is 10. The total distance is 15, and the total cost is 35. For the second path, the distance between **A** and **C** is 5 units, and the cost is 5 units. The distance between **C** and **D** is 5 units, and the cost is 5 units. Finally, the distance between **D** and **E** is 10 units, and the cost is 10 units. The shortest path between **A** and **E** in terms of distance is going through **B** (15 units of distance against 20 through **C** and **D**). However, the shortest path between **A** and **E** in terms of cost is going through **C** and **D** (20 units against 35 through **B**). **B** partakes a shortest path if the constraint is the distance. **C** and **D** partake a shortest path if the constraint is the cost.

Every node partaking a shortest path somehow controls the information flow within the network. The shortest paths taken equals more control over the information flow. These types of nodes work as gatekeepers within the network. Betweenness centrality counts the number of shortest paths a node partakes, and hence, how much it can control the information flow. The constraint in social network is usually the strength of the relationship. So, it is a maximization problem. The higher the sum of the link weights, the better. Therefore, the higher the betweenness centrality, the more the node controls the information flow within the network.

Closeness

Closeness centrality accounts for the average shortest path from a node to the rest of the nodes within the network (or subnetwork). The concept of shortest path is the same as described to the betweenness centrality. If a node has a small measure for the average shortest path to all other nodes in the network (or in its subnetwork), it means that this node is close to every other node. Analogous to the betweenness centrality, the constraint is the link weight, and it is a maximization problem. The stronger the connections, the better. This calculation can be understood as the longest path because the relevance here is based on the strength of the relationships. Based on that, the higher the closeness the closer the node is to the others in its network.

The business concept of the closeness centrality is assigned to the speed a node can spread some information throughout the network. It is the gossip person. A node with high closeness centrality can spread a message throughout the network faster than other nodes. This information can be used to target customers to spread out a message about a new product. They would be targeted in a product adoption campaign. Like degree centrality where in- and out- designations are required to indicate incoming and outgoing connections in directed networks, those designations are also used in closeness centrality measures when the network is directed. Then, for directed networks we calculate closeness-in and closeness-out. The overall closeness is the sum of closeness-in and closeness-out.

Hub and Authority

Hub and authority centrality were originally developed to rank the importance of web pages. Certain web pages are important if they point to many important pages. These web pages are called hubs. Other web pages are important because they are pointed to by many important web pages. These web pages are called authorities. The same concept can be extended to any type of network to rank or measure the importance of the node by the number of outgoing or incoming connections from and to other important nodes. Because of the distinction between incoming and outgoing connections, hub and authority are computed just for directed networks. A good hub node is the one that points to many good authority nodes, and a good authority node is the one that is pointed to by many good hub nodes.

This concept is easily understandable when applied to a co-authorship network. A paper that cites many good authority papers has a high hub score, and a paper that is referenced by many other influential papers has a high authority score. We can also think about the authority score as a person who is the expert in some subject matter. Lots of people consult and refer to them. They are an authority, as we say. The concept of hub score can be easily understandable when we think about airports. The hub of Delta Air Lines in the United States is Atlanta. Most of the flights to international or long-haul destinations depart from Atlanta.

Eigenvector

Eigenvector centrality considers the importance or strength of the connections of the adjacent nodes. However, it does not weigh all the neighbors equally. Using eigenvector centrality, connections to more important nodes contribute more to a node's centrality score than connections to less important nodes. An eigenvector measures the importance of a node inside the network. Relative scores for all nodes are computed based on their connections, considering both the frequency and strength of the relationship. The eigenvector is assigned to a recursive algorithm that calculates the importance of a node, considering the importance of all nodes and all connections within the network.

PageRank

PageRank centrality, named after Google's cofounder Larry Page, was originally proposed for the application of ranking web pages in a search engine. The PageRank algorithm models the stationary distribution of a Markov process, assuming that each node is a web page, and each link is a hyperlink from one page to another. A web surfer might choose a random link on every page or might jump to a random page on the whole web with some probability. In PageRank centrality, being pointed to by a particularly important node is buffered if that important node contains many outbound links. For example, if an especially important web page contains an exceptionally large number of URL (Uniform Resource Locator) links, the endorsement value for each outbound web page does not carry as much value compared to the same important web page pointing to only a few selected URL links.

One of the main differences that distinguishes PageRank centrality from other centrality measures is that the importance of the node can vary based on the number of outbound links an adjacent node has. In other words, there is an element of exclusivity implied in the importance of that node when pointed to by other important nodes. Because the distinction of the inbound and outbound links to the adjacent nodes, the PageRank centrality is computed just for directed networks.

Use Case: Social Network Analysis to Reduce Churn in Telecommunications

A telecommunications company has information about when customers get connected to each other through several types of relations like calls, text messages, and multimedia messages. This company also knows the frequency and recency of customers' connections to each other. Based on that information, this telecommunications company can use the connected data to build a network of customers. In this network (Figure 8.3), the customers are the nodes, and their connections are the links. The weights of the nodes are based on their average revenue per user (ARPU). The weights of the links are based on the weighted sum of their connections, considering different weights or values for calls (including frequency and duration), text messages (frequency), and multimedia messages (frequency and size).

The main goal of the network analysis for this case is to detect communities within the entire network and compute the network metrics based on the communities identified. The network metrics are standardized and summarized to create a single centrality. The node(s) with the highest centrality within each community are the leader(s). A retention campaign can be launched based on these leaders to avoid their churn and assuming that avoiding their churn the company is avoiding their peers' churn as well.

A control group was defined to evaluate the performance of this retention campaign. The outcomes of the control group can also help the company estimate the impact of targeting the right customers considering their influential factor. For example, the regular customers connect to 36 other subscribers

Figure 8.3: Telecommunications Network

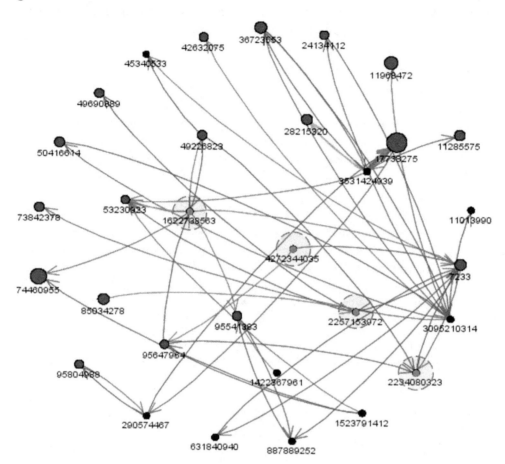

on average, of which 29 belong to the same carrier (same company – not the competitor). Once they decided to leave, other peers decided to leave afterward. On average, each one of these customers was able to influence 0.87% of their peers. On the other hand, the influencer customers connect to 55 other subscribers on average, of which 45 belong to the same carrier. On average, each one of these influencer customers was able to influence 5.94% of their connections. This influential impact is about seven times greater than the regular customer, when they decide to make churn. These figures would show the company the importance of reaching out to the influencer customers first or with a more aggressive retention campaign.

Figure 8.4 shows that a regular customer relates to 36 other customers; 29 belong to the same carrier. On average, a regular customer can influence less than 1% of their peers to leave the company in case they make churn. An influencer customer relates to 55 other customers, 45 of which belong to the same carrier. On average, they can influence 6% of their peers to follow in case they decide to make churn.

Figure 8.4: Customer Influence Within the Network

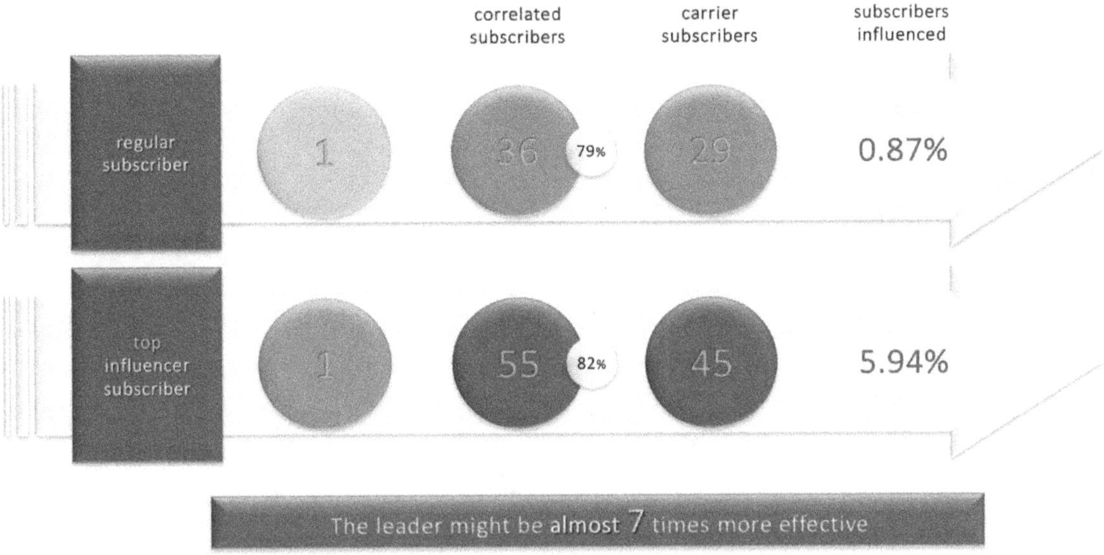

Network Optimization
====================

Another type of analysis of network or graph structure is network optimization. Network optimization considers a set of optimization techniques used to evaluate graph structures, identify subgraphs, evaluate network performance, and search for optimal solutions in network topologies. For example, data scientists can use network optimization to study the traffic flows within geographic areas. In such studies, the nodes are the geographic locations, and the links are the people traveling between these locations. The weight of the locations could be the population, and the weight of the link could the number of people traveling. This information can be truly relevant in helping government agencies plan public transportation systems, road networks, and emergency evacuation plans, among many other applications. Health agencies can also use this set of techniques to study the vectors of infectious diseases. Communications and utility companies can use these graph data and algorithms to better plan their physical distribution networks and forecast the usage. Financial institutions can use network optimization to fight against money laundering, and governments can use it to understand the major mechanisms of corruption. There is an endless number of applications that can be deployed by using network analysis and network optimization. The important concept of network science is that everything is connected and somehow can be elaborated as a network problem.

Network Algorithms

Network optimization considers graph theory and optimization algorithms that can augment and generalize mathematical optimization approaches. Many practical applications of optimization depend on an underlying network. For example, retailers face the problem of shipping goods from warehouses

Figure 8.5: Network Optimization: Traveling Salesman Problem – Optimal Tour

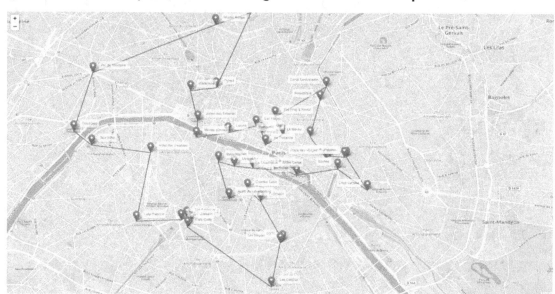

to stores in a distribution network to satisfy demand at minimum cost. Commuters choose routes in a road network to travel from home to work in the shortest amount of time or traveling the smallest distance. Financial institutions search for sequences and paths that can be associated with illicit transactions.

The following network algorithms aim to identify subgraphs within the entire network or the best way to perform some activity considering the network structure.

Clique

A clique is a subgraph that is a complete graph. Every node in a clique is connected to every other node in that clique. It is a sequence of nodes such that, from each of the nodes within the graph, there is a link to the next node in the sequence. Finding cliques in graphs has applications in many industries such as bioinformatics, social networks, electrical engineering, and chemistry. A trivial example of a clique within a network is a group of nodes that are connected to each other. For example, subscriber A is connected to B, C, and D. Subscriber B is also connected to C and D, and subscribers C and D are connected. Links in a clique would be like this: A–B, A–C, A–D, B–C, B–D, C–D. Cliques can be quite useful in finding some groups of interest within a network.

Cycle

A path in a graph is a sequence of nodes such that, from each of the nodes within the graph, there is a link to the next node in the sequence. A cycle is a path in which the start node is the same as the end node. The method is aimed at directed graphs. For undirected graphs, each link represents two directed links. A trivial cycle is $A \rightarrow B \rightarrow C \rightarrow A$.

Linear Assignment

A linear assignment problem is a fundamental problem in combinatorial optimization that involves assigning workers to tasks at minimal costs. In graph theory, the linear assignment is like finding minimum weight matching in a weighted bipartite graph. A bipartite graph is a graph where nodes can be divided into two disjointed sets S and T (workers and tasks) such that every link connects a node in S to a node in T. For example, a relay swimming team has 4 swimmers with respective best times for the different styles: butterfly, breaststroke, backstroke, and freestyle. Forming the best team is not just a matter of getting the fastest swimmer in each style, but to combine the swimmers to minimize the overall time.

Minimum-cost Network

Minimum-cost network flow is a fundamental problem in network analysis that involves sending flow over a network at minimal cost. Minimal cost here is the objective function. The objective function can also be a minimal time or a maximal quantity. It depends on the problem data scientists are studying. The network is created based on the links or connecting nodes. Nodes can be, for example, warehouses and stores. Links can be roads. Each link in the network has an associated cost. It can also have an associated minimum and a maximum capacity.

For example, to flow products between warehouses and stores, some roads might have a maximum capacity of goods to be transported. Some roads might have a toll, which represent the cost. If transporting through a third-party company, a minimum number of trips might be required. The stores have demands and the warehouses have supply capacities. The decision variables define the amount of flow to be sent between the nodes or the products to be transported between the warehouses and the stores. This problem can be modeled as a linear problem, and the goal is to optimize the number of goods to be flowed at the minimum cost.

Minimum Spanning Tree

The minimum spanning tree problem is aimed at identifying the minimum number of nodes that connect the entire network. Consider a general network. This network has multiple nodes and links. There is a reachability assigned to this network considering all the connections between the nodes through the available links. The minimum spanning algorithm aims to find the minimum set of links that keep the same reachability of the original network. In other words, this minimum set of links allows the origin nodes to reach the same destination nodes by using this small new set of connections. A single graph can have many different spanning trees.

This problem is useful in real-life applications such as the design of networks like telecommunications, computers, transportation systems, water supply, and electrical grids, among others. A spanning tree of a connected undirected graph is a subgraph, which is a tree that connects all the nodes together. Given weights on the links, a minimum spanning tree is a spanning tree with weights less than or equal to the weight of every other spanning tree. Any undirected graph has a minimum spanning forest, which is a union of a minimum spanning tree and its connected components.

Path Enumeration

A path in a graph is a sequence of nodes such that from each of the nodes within the graph there is a link to the next node in the sequence. An elementary path is a path in which no node is visited more than once. A path between two nodes in a graph is a path that starts at the origin node and ends at the destination node. The origin node is called the source node, and the destination node is called the sink node. A network optimization algorithm for path enumeration usually searches for all pairs of nodes in the network. That is, it finds all paths for each combination of source nodes and sink nodes. This information is useful to understand the level of reachability of the network and all routes between nodes considering the available links within the network.

Shortest Path

The shortest path problem aims to find the path between two nodes in a network where the sum of the weights of its links is minimized. In other words, the algorithm searches the paths that satisfy the objective function. The objective function can be defined as minimizing the time, minimizing the distance, minimizing the cost, maximizing the strength of the relationship, and so on. Real-life problems that can be addressed by the shortest path are public roads networks, public transportation systems, traffic planning, sequence of choices in decision making, and minimum delay in telecommunications operations, among others.

A shortest path between two nodes in a network is the path that starts at the origin node and ends at the destination node with the lowest total link weight. The link weight here can be the time to go from one node to another, the distance, the cost, or the strength of the relationship between persons, among other types. The origin node is referred to as the source node and the destination node is referred to as the sink node.

Transitive Closure

Transitive closure of a network is a graph such that for all nodes in the network there is a link between them. The transitive closure of a network can help efficiently answer questions about reachability. Suppose you want to find out whether you can get from one node (the origin location) to another specific node (the destination location), considering the available links within the network. Data scientists can use the transitive closure network optimization algorithm to calculate all connections within the network (the existing reachability of the network) and simply check for the existence of links between a particular pair of nodes (or a route between a specific pair of locations – origin and destination). Transitive closure has many applications in real-life problems, including speeding up the processing of structured query languages, which are often used in databases.

Traveling Salesman Problem

The traveling salesman problem (TSP) aims to find the minimum-cost tour in a network. A path in a network is a sequence of nodes linked to each other. An elementary cycle is a path where the start node and the end node are the same, and no node is visited more than once in that sequence. A tour is an elementary cycle where every node is visited. The traveling salesman problem aims to find the tour with minimum total cost. Again, the minimum cost here can be the actual cost, the distance, or the time. The traveling salesman problem can answer questions about optimal tours. For example, given a list of locations and the distances between each pair of locations, what is the shortest possible route that visits every location in my tour exactly once, starting from one location and returning to it? The traveling salesman problem has applications in many areas like planning logistics routing, manufacturing, and supply chain, among others.

Pattern Match

Imagine two different networks: G, the main network, and Q, the query network. Subgraph isomorphism is the problem of finding all subgraphs of G that are isomorphic to Q, which means that all subgraphs that have the same topology as graph Q. Pattern matching addresses the analogous problem in the presence of node and link attributes. It is the problem of finding all subgraphs of the network G isomorphic to the subnetwork Q such that all node and link attributes defined in Q are preserved in subnetworks of Q under the isomorphism map. Subgraph isomorphism and pattern matching have applications in many areas, including social network analysis, fraud detection, pattern recognition, data mining, chemistry, and biology.

Use Case: Smart Cities – Improving Commuting Routes

A splendid example of network optimization is smart cities. Data scientists can use network optimization algorithms to understand human mobility behavior and then plan the commuting routes in metropolitan areas. A key data source for this type of project is the mobile data provided by telecommunications companies. Subscribers' mobile data can be anonymized and aggregated for network optimization. Spatiotemporal information – geolocation and time – can be used to create a network of subscribers' mobility. A set of network algorithms can be used to describe the network topology and its main characteristics and then to improve some specific aspects of the network.

This project investigates the daily commuting routes that workers in Rio de Janeiro take. The mobile data – geolocation and time – was provided by a mobile carrier, and it was anonymized and aggregated to protect customers' privacy.

Based on the most frequent cell tower each subscriber uses overnight and during business hours, data scientists can infer their presumed domiciles and their presumed workplaces.

Figure 8.6: Cell Towers Based on Frequency of Visitors

Figure 8.7: Cell Tower Based on Frequency of Visitors During Work Time and Nighttime

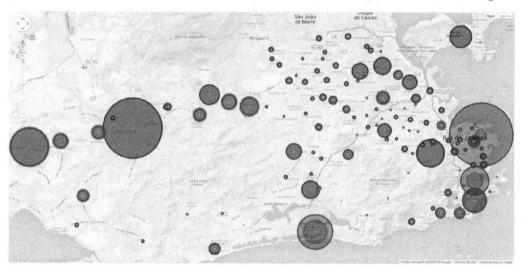

Every time a subscriber moves from one cell tower to another in a specific time frame (every hour for example), a movement vector is created. All movement vectors from one cell tower to another in a time frame can be summarized to define the overall displacements between cell towers. The cell tower is the lower granularity in terms of geolocation. Vectors of movements can be summarized to determine displacements between pairs of locations, such as neighborhoods, administrative sectors, counties, cities, and so on. The major vectors of movements can be selected to estimate the overall commuting routes. The number of people traveling throughout these major commuting routes can

Figure 8.8: Most Frequent Commuting Paths

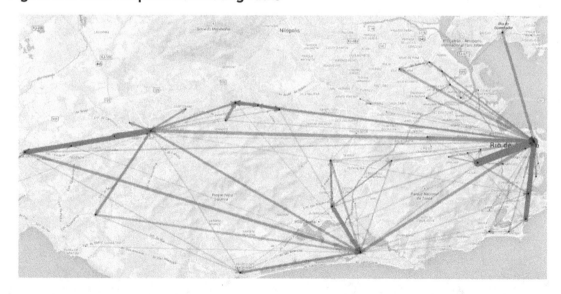

be extrapolated by the number of subscribers, the mobile penetration in the area, and the population of the area. Based on that, local government agencies can better plan for transportation systems and traffic routes by knowing the number of people traveling between all pairs of locations and the trajectories that they need to perform to arrive at their destinations.

Summary

This chapter briefly introduced concepts about network analysis and network optimization. These two topics are under a major discipline called network science. Network science studies patterns of connections usually assigned to complex networks. Anything could be a link to connect entities: a call, a letter, a flight, a bank transfer, a vote, an email, a claim, virtually anything. Anything can be an entity: a bank account, a person, a company, a country, a cell phone, virtually anything. Network science consists of a set of algorithms and methods to create a network from raw data, evaluate the network, compute metrics, and find hidden patterns. Network analysis algorithms evaluate the role and the importance of the nodes and links within a network. Network optimization algorithms are used to improve performance in a specific network, based on a set of constraints and resources. It aims to ensure an optimal network design with the lowest cost structure and maximal flow.

Chapter 9: Model Assessment and Model Deployment

Chapter Overview

At this point, we have covered most of the data science methods, including statistical approaches and machine learning techniques. We also covered supervised and unsupervised models.

In this concluding chapter, we are getting closer to the end of the analytical life cycle. Remember in Chapter 1 when we first introduced the analytical life cycle: the problem definition phase, data preparation, model training, model assessment, and deployment. Now it is time to evaluate the model's results and select the champion model that we are going to put in production to support a business action.

We also are going to learn about the most important phase for the company. It might not be the most important phase for the data scientist since they probably enjoy the model development. However, if the company is waiting for the model to support some business actions, the model deployment phase might be the most important phase of the analytical life cycle for the company.

The main goals for this chapter are the following:

- Describe the several types of fit statistics available for assessing the model's performance.
- Explain how to use fit statistics to evaluate the results of multiple models according to the business goals.
- Explain how to use results generated by analytical models to create deployable business actions.
- Describe different methods for deploying analytical models into production to solve business problems.
- Describe techniques to score models in batch and real time.
- Describe methods to manage and monitor models in production.

Methods to Evaluate Model Performance

The main goal of analytical models, particularly supervised models, is generalization. *Generalization* is the performance of the model on new data. It is important that we keep the good model performance that

Figure 9.1: Aspects to Evaluate the Best Model

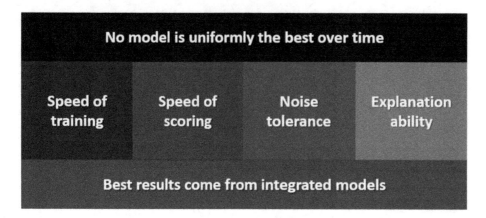

we observed during the training and validation phases when the model is deployed during the production stage. This is important because the model in production is the one that will support a business action, a sales campaign, a retention offering, or fraud detection, among others. The model in production needs to provide satisfactory performance, a reasonable accuracy when estimating or classifying a target, or a reasonable fit when scoring a new case. The model was trained and validated based on past data, but the main goal here is to use it on new data where the new observations have no target if the model is supervised or the observation was not in the data used to create the insights if the model is unsupervised. For example, a supervised model such as a churn model can be deployed to score and classify customers based on the probability of churn, and therefore to receive an incentive to stay if the probability is above a certain threshold. An example of an unsupervised model in deployment is a customer segmentation model that scores and classifies new customers into one of the segments previously created.

There are many ways to evaluate the model's performance. In addition to different fit statistics, there is a vital component: the business need. Depending on the business scenario, the model can be evaluated based on its accuracy such as how good it is in predicting the new observation, but that model can also be assessed based on the time it takes to be trained, the time it takes to score new cases, the ability to handle noise within the data, and finally, the capacity to be interpreted.

Speed of Training

The speed of training is important in some fraud detection models. The fraud behavior might change so quickly that the model eventually drifts too early. When that happens, the model in production might need to be retrained. In the worst-case scenario, the model might need to be re-created. In both cases, the time to train or retrain is completely relevant.

Speed of Scoring

The speed of scoring is especially important in real-time applications. For example, in credit risk scoring, customers call the call center or go straight to the company's web page to apply for a personal loan. The business application, either the CRM (Customer Relationship Management) on the call center or the web page application, needs to collect the customer information and in a matter of seconds reply to the response. The input data is collected by the business application, and the model previously trained is applied to the new data to score the credit risk for that loan application.

Some models are fast to train, like regressions and decision trees. Some other models can take longer to train on big data, like neural networks and gradient boosting. On the other side, some models can be extremely fast to score, like regressions and neural networks, as the scoring is just a matter of applying the coefficients or weights to the new data. The speed of the training in cases where the behavior changes quickly, or the speed of the scoring in cases where the time response is required to be low, can be the most important dimension to evaluate and select the best model to put in production.

Business Knowledge

Depending on the type of unsupervised model, there are some statistical measures that data scientists can use to evaluate the results. For example, in clustering, you can use the cubic clustering criterion to evaluate the possible number of clusters based on the minimum variance in each set of groups. In network analysis, data scientists can use modularity to evaluate the best distribution of communities detected. In some other clustering techniques, data scientists can use statistical measures to evaluate the proportion of the variability in distinct groups. However, in most of the cases of unsupervised models, the business knowledge and the experience of the data scientist can be the key. As unsupervised methods are more open in terms of results, the assessment of the model results needs to rely on the business knowledge, on how the results are applied as business actions, and on what the company expects as outcomes.

Fit Statistics

For supervised models, on the other hand, data scientists have a baseline that is a benchmark to evaluate how good the models are that they developed. The baseline is the known target used to train the models. All supervised models are trained on past data where the outcome is known. Once the model is trained, data scientists can validate the models based on the results of these models on different data sets. The main goal on this approach is to select the model that can best generalize the predictions in different observations.

The fit statistics used to evaluate supervised models really depend on the type of model. If the supervised model is a classification model and the target is a categorical variable, the outcome is a decision. Then, the fit statistics used to assess different models' results can be accuracy, misclassification, profit, or loss. The model with the lowest misclassification on new data is the champion model. If the supervised model is a ranking model where the output value is less important than the order of these values, the outcome is a ranking. Then, the fit statistics used to assess different models' results can be the ROC index or Gini coefficient. The champion model is the one that can best generalize the ranking on new data or the model that presents the highest ROC index. Finally, if the supervised model is an estimation model, where the target is a continuous variable, the outcome is an estimate. Then, the fit statistics used to assess different models' results can be average square error, root mean square error, Schwartz, Bayesian, or Akaike information criterion, or any

Figure 9.2: Types of Predictions and Fit Statistics

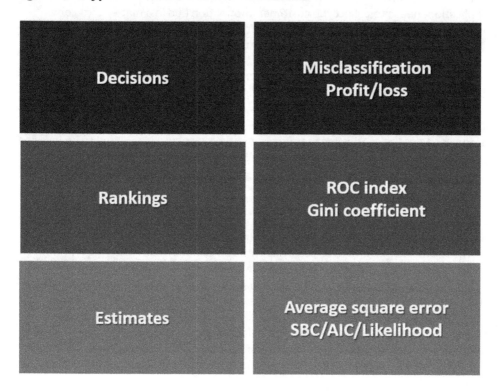

statistics based on a likelihood function. The model with the lowest error function on the new data is the champion model.

Data Splitting

The model trained on a specific set of data in a specific time frame is trained to identify the correlation between the input variables to the target variable. This model will be applied to new data or future data to support a business action initiated by the company. For example, a company is experiencing a rise in the churn rate. The data science team can collect past data about the churn event and train supervised models to understand and identify what is correlated to the churn. What type of customers are more likely to make churn? What customers' characteristics are more relevant when analyzing the churn?

Once a set of models is trained, data scientists need to evaluate the models to select the best one—the one that they deploy in production to support a retention campaign. However, this retention campaign will run on a future time frame. The variables used to train the models are the same, but not their values. A period has passed, and these values might have changed. The data the model scores is different from the training data. To assess whether the model might fail on this new data, data scientists evaluate the model based on a different data set than the one on which the model

Figure 9.3: Model Generalization: Train, Validate, and Test

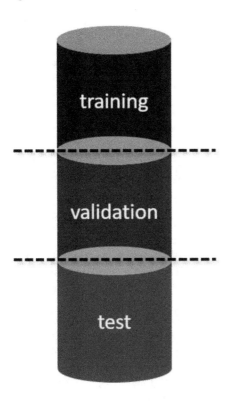

was trained. This is to "simulate" the real-world scenario, when the champion model scores different data than it was trained on. This different data used to evaluate the best model during the training can be two different data sets called validation data and test data. The best model will be the one that performs best on the validation and test data sets, not on the training data set. Data scientists can use the results of the models on the validation data set, the test data set, or the average results on both validation and test data sets to select the champion model. When this champion model is deployed in production to support the campaign, the data scientists can expect that the model will perform well because they selected the model that generalized the best, using different data sets to train and select the champion model. In some cases, the test data can be an out-of-time sample to consider a future time frame of events to simulate what will happen when the model scores new or future observations.

K-fold Cross-validation

Splitting the data into training, validation, and test (The test data set is not really required but depends on business needs and regulations.) is a highly effective method to train and generalize a supervised model before getting it into production. Data splitting is a simple but costly technique. When the data set is too small, the performance measures might be unreliable because of high variability between

Figure 9.4: Cross-validation

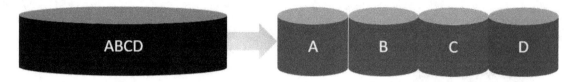

the train and validation data sets. A good approach to mitigate this problem of small data sets is cross-validation.

K-fold cross-validation shuffles the original data set into training and validation. Multiple runs can be defined for this shuffle, or multiple folds can be determined to split the original data set into training and validation. For example, in a four-fold cross-validation, the data is split into four equal sets. The first run of the training is based on three quarters of the original data set (folds A, B, and C), and the validation is based on the last quarter (fold D). A new run uses folds A, B, and D to train and validate based on C. The process continues until all folds have been used for the training and the validation steps. The four assessments are then averaged. In this way, all the data are used for both training and validation.

There are two options for the best model in cross-validation. The first one considers a single model for each step. For example, a model is trained on the first step based on the folds A, B and C, and validate based on the fold D. This model is then applied to all other steps, the other three shuffles. The model that performs best in all folds or steps is the champion. Some data scientists use the entire data set to fit the champion model but use the cross-validation results as the assessment statistics.

The second approach is the ensemble model. For every step, a model is trained like in the first approach described before, for example, training on A, B, and C and validating on D. Another model is trained on the second step, training on A, B, and D and validating on C, and so on. In this case, there are four models trained: one for each run of the cross-validation. Because the results are averaged, the models are an ensemble. The final model is the ensemble model of all four models trained during the cross-validation.

Goodness-of-fit Statistics

The method to assess the models' performance depends on the scale of the target variable. If the target variable is categorical, then some of the common methods are classification tables, ROC curves, gains, and lift charts. Several of the goodness-of-fit statistics can be used to evaluate the models' performance. Some of the most common goodness-of-fit statistics are the misclassification error rate and information criteria, such as the AIC (Akaike information criterion) and SBC (Schwarz Bayesian information criterion).

If the target variable is continuous, a plot of the observed versus predicted averages by decile would be useful. Goodness-of-fit statistics are commonly based on some error functions, like average squared error, root mean square error, log error, and the adjusted R-square.

Figure 9.5: Confusion Matrix

	Observed 1	Observed 0
Predicted 1	True-Positive (TP)	False-Positive (FP)
Predicted 0	False-Negative (FN)	True-Negative (TN)

Confusion Matrix

One of the most widely used assessment tools for supervised classification models when the target is a binary variable is the confusion matrix. The confusion matrix is a crosstabulation of the actual (observed event) and predicted classes (classified event). It quantifies when the supervised model correctly classifies or misclassifies the event. For the confusion matrix, the event of interest is often called positive whether the event is unfavorable like fraud, churn, or default, or the event is favorable like purchase, acquisition, cross-sell, or up-sell. A nonevent is called negative.

If the case is predicted to be positive (1) and it is observed to be positive (1), then this case is a true-positive (TP). If the case is predicted to be positive (1) and it is observed to be negative (0), then this case is a false-positive (FP). If the case is predicted to be negative (0) and it is observed to be positive (1), then this case is a false-negative (FN). Finally, if the case is predicted to be negative (0) and it is observed to be negative (0), then this case is a true-negative (TN).

All the cases are allocated to classes based on cutoff values of the predicted probability. The steps include the following:

1. Estimate the predicted probability of class 1 for each case.
2. Choose a cutoff probability (often the event proportion in the sample).
3. Assign the cases to class 1 if their estimated predicted probability exceeds the cutoff. Otherwise, assign the cases to class 0.

Based on the decision (according to the predicted probability cutoff), the simplest performance statistic from the classification table is accuracy. Accuracy can be calculated as the following equation:

$$ACC = \frac{TP + TN}{TP + FP + FN + TN}$$

The error rate, also called the misclassification rate, is the complement of the accuracy rate:

$$ERR = 1 - ACC = \frac{FP + FN}{TP + FP + FN + TN}$$

Different cutoffs can produce different allocations for the decision class and then different classification table values. A change in the cutoff value produces different performance statistics like accuracy or misclassification. Depending on the event distribution, increasing the cutoff might increase or decrease the accuracy and decrease or increase the error. A general recommendation for a cutoff is the proportion of events in the training data set. This information is extremely sensitive particularly when classifying rare events. For example, assume that a data scientist is training a model to detect fraud. In most cases, fraud is a rare event. Let us say that for this specific case, fraud represents 1% of all the transactions. If the final model says that there is no fraud ever (that is, predicts all cases as nonfraud), the model is correct in 99% of the cases, and its overall accuracy would be 99%. A model performance with 1% of misclassification can be considered as a wonderful performance. However, the company would miss all the fraud cases.

Because of that, there are two specialized measures to evaluate the performance of a supervised classification model. The first one is sensitivity, also known as recall or the true-positive rate, which can be calculated based on the following equation:

$$Sensitivity = Recall = TPR = \frac{TP}{P} = \frac{TP}{TP + FN}$$

The second specialized measure is specificity, also known as selectivity or the true-negative rate. Specificity is the proportion of nonevents that were predicted to have a nonevent response. It can be calculated as the complement of the sensitivity, based on the following equation:

$$Specificity = Selectivity = TNR = \frac{TN}{N} = \frac{TN}{TN + FP}$$

The context of the problem determines which of these measures should be the primary concern of a data scientist. For example, in a retention campaign, the data scientist should be concerned about the false-negative rate (FNR or FN/FN+TP). The false-negative cases (FN) would give the number of customers not contacted during a retention campaign that would be lost. The false-positive cases (FP) would not be the worst case. The company would be contacting customers who are not willing to make churn, and they would eventually receive an incentive to stay. That would work as a marketing or relationship campaign. However, losing customers because they were not identified as churners during a retention campaign would be bad.

On the other hand, in a fraud detection application, data scientists should be more concerned with the false-positive rate (FPR). The false-positive cases (FP) would give us the number of good customers who were identified as possible fraudsters. Of course, the false-negative cases (FN) would also be bad. They would represent the number of possible fraudsters not identified by the model. These fraudsters would not be identified as committing fraud events against the company. However, the false-positive cases might represent good customers being blocked or having their services cancelled by mistake. The negative impact of the false-positive cases might be higher than the fraud itself in terms of the company's image, or good customers suing the company, or influential customers spreading a bad message about the company's services.

ROC Curve

Another common method to evaluate a model's performance for classification models is the Receiver Operating Characteristic (ROC) curve. The ROC curve is a graphical display that gives a measure of the predictive accuracy of a supervised classification model with a binary response. The ROC curve displays the sensitivity or the true-positive rate against 1-specificity or the false-positive rate. The ROC curve displays this relationship based on a range of different cutoffs.

The extremes (0,0) and (1,1) represent cutoffs of 1.0 and 0.0, respectively. If a data scientist uses the rule that all observations with a predicted probability of 1 are classified as events, none of the event observations are correctly classified, but all the nonevent observations are correctly classified (sensitivity=0, specificity=1). If a data scientist uses a rule that all observations with a predicted probability of 0 or higher are classified as events, all the events are correctly classified, but none of the nonevents are correctly classified (sensitivity=1, specificity=0). The horizontal axis is the false-positive rate, or 1–specificity, so the ROC curve starts at point (0,0), and the vertical axis is the true-positive rate, or the sensitivity, so the ROC curve ends at point (1,1).

For a supervised classification model with high predictive accuracy, the ROC curve rises quickly (sensitivity increases rapidly, specificity stays at 1). In such cases, the area under the curve is large. In contrast, the ROC curve rises slowly and has a smaller area under the curve for supervised classification models with low predictive accuracy. The area under the curve is not affected by the cutoffs the way that misclassification and accuracy are affected.

An ROC curve that rises at 45° is a poor model. It represents a random allocation of cases to the classes and is the ROC curve for the baseline model.

Figure 9.6: ROC Curve

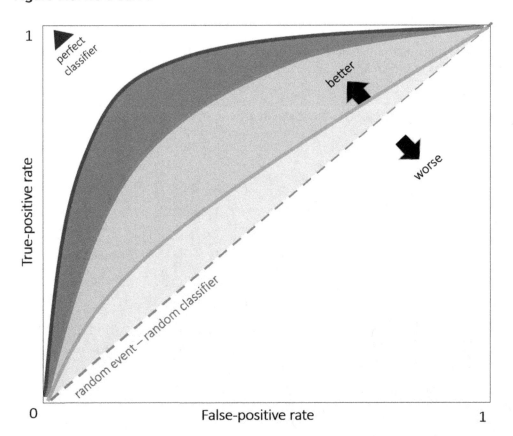

Cumulative Percentile Hits Chart

The depth of a classification rule is the total proportion of cases that were allocated to class 1. The cumulative percentile hits chart displays the positive predicted value and depth considering the overall population. (See Figure 9.7.) For example, assuming a 5% response rate for a sales campaign, at 5% of the population, we can expect to observe a 5% response rate. At 10% of the population, we can expect to observe the same 5% response rate and so on.

When data scientists compare the model's performance with the random event, they can observe how well the models perform. For example, in Figure 9.8, ranking the population based on the predicted probability of the event, or in our case here, to purchase the product offered during the sales campaign, we can observe that at the top 5% of the population, the response rate reaches 21%. A simple comparison in terms of model performance would be like this: if there is no model, a sales campaign would randomly select 5% of the population to offer a specific product. This 5% of the population could be based on a limited budget to run the campaign. Based on the average response rate for this campaign, the company could expect a response rate at 5%. In other words, just 5% of the customers contacted would purchase the product offered. By using the supervised classification model and contacting only 5% of the total population, the same campaign would target the top 5% of the customers in terms of the predicted probability, or the customers more likely to purchase that product. By using the model, the company would observe a response rate of 21%, or 21% of the customers contacted with the offer would purchase the product.

The data scientist could show that the model is four times better than the random event. In other words, the supervised classification model could increase the response rate for this sales campaign by four times.

Figure 9.7: Percentage Cumulative Hits – Random Events

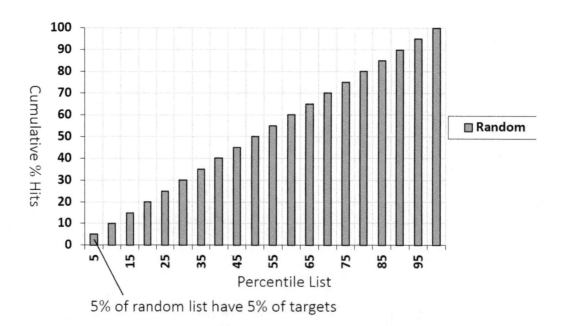

5% of random list have 5% of targets

Figure 9.8: Percentage Cumulative Hits - Model versus Random Events

5% of random list have 5% of targets
5% of model ranked list have 21% of targets (CPH(5%,model)=21%)

Lift Chart

The lift is calculated as the positive predicted value over the overall response rate. So, for a given depth, there are more responders targeted by the model than by random chance. The graph shows that at the fifth percentile the lift is 4.2 (21% of responders by using the model divided by the 5% of responders by not using the model) (random event).

Figure 9.9: Lift Chart

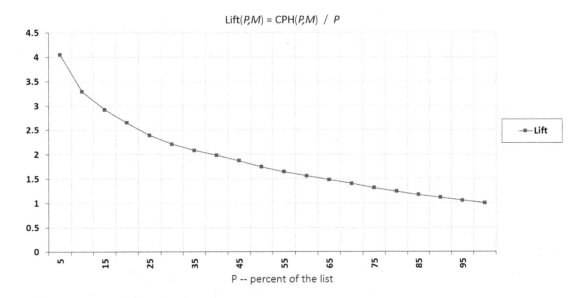

Lift charts are widely used in database marketing to decide how deep in a customer population the company should go for a campaign. The simplest way to construct a lift curve is to sort and bin the predicted posterior probabilities. The lift chart can be augmented with revenue and cost information.

The lift chart can also be used to evaluate and select the model to be deployed in supporting a specific campaign. For example, if a company has a limited budget for a sales campaign and can contact only 20% of the total population, then the data scientist can select the model that performs best for the top 20% of the population based on the predicted probability for the target (purchase the product) instead of selecting the model that performs best considering the entire population.

Model Evaluation

Data scientists are recommended to try multiple models and consider different methods and distinct algorithms. All the techniques are data-dependent, and trying different techniques, multiple modeling approaches, and data preparation approaches are always helpful. Depending on the data variability and the relationship between the input variables and the target, a regression might fit better than a neural network. Often, if the relationship between the inputs and the target is simple, a regression model, a decision tree, and a support vector machine can fit properly and end up as the best model. In these cases, there is no reason to have a complex model like a neural network or an ensemble model like a forest or gradient boosting model if a simple and interpretable model like a regression model and a decision tree can be used. However, when the relationship between the inputs and the target is complex, more complex models like neural network and ensemble of trees usually fit better. The drawback is the lack of interpretability.

Performance

Data scientists always evaluate the models to select and to put in production according to multiple dimensions. The first dimension is performance. Performance can be measured by using different fit statistics, and it relies on the type of the target. The target might be a class, which implies a decision, a ranking, or an estimation. Each one of these types of targets can be evaluated by different fit statistics.

Data scientists can work with the business department to understand what fit statistics are most relevant to the business. For example, the most important fit statistic might be the overall accuracy. In that case, the data scientist can select the model with the minimum misclassification error. The business event can be extremely sensitive, like fraud, and the false-positive can represent a huge risk for the business. The data scientist can then evaluate the best model based on the minimum false-positive rate. Some other business events like churn might be impacted by the false-negative rate, which misses the customers who are likely to make churn. Finally, the model can be evaluated by looking at the accuracy on a specific percentile of the population. A marketing campaign might be targeting only a small percentage of the population. The data scientist can evaluate the champion model based on the performance of the models just on the top percentile associated with the campaign.

Time to Train Model

The second dimension that a data scientist can use to evaluate models is time to train the model. For example, in fraud, the model might drift early and should be trained – or retrained – quite often. Based on that, the time to train the model is crucial. In credit risk, the model needs to score in real time when the customer applies for a loan. In some projects, the scoring should be embedded in the transactional system. The model should be converted into a set of rules or into an equation. Models like decision tree and regression are good candidates for this type of translation. Sometimes the model scoring is consumed by another application as an application programming interface (API). The model selected as a champion should provide this type of access. All these types of evaluation go far beyond just the accuracy of the models.

Interpretability

Interpretability is also a crucial measure for some business events. Particularly in credit scoring, the company might be requested to provide an explanation about the models results, such as the reason a customer had a loan denied. For these cases, interpretability is an essential measure to evaluate the champion model that will be put into production.

Generalizability

Generalization is also very important to evaluate the models that are going to be deployed. Based on that, the candidate models can be assessed based on an honest test, an out-of-time data set. Sometimes the data scientist needs to evaluate the best model based on a combination of some of the characteristics listed above.

Model Deployment

The analytical life cycle comprises the model development (including the training of multiple and different models), the model comparison (considering the evaluation of different models based on the business needs), the model selection (taking into account the business requirements – speed, interpretation, etc.) and the model deployment (considering how the model needs to be put in production – batch, real time, API, etc.).

Figure 9.10: Analytical Modeling Life Cycle

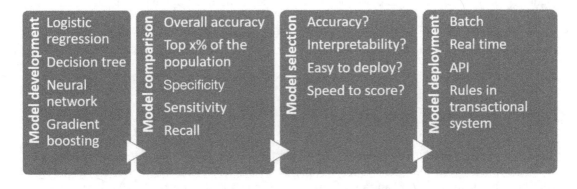

Challenger Models

The analytical life cycle should involve model operationalization. As highlighted before, it is the most important phase for the company when the model selected as the champion model supports a specific business action. It is important to have challenger models as well. The challenger model can be one of the models evaluated during the assessment that was not selected as the champion model, perhaps because the overall accuracy, time to score, or lack of interpretability. Regardless of the reason they were not selected, challenger models are good models that can replace the champion model in case something happens. For example, let us say that customer behavior changes over time and the model performance decays. The champion model is not generalizing as expected, and before another model is trained, one of the challenger models can be deployed to replace the current champion. This model replacement would save a significant amount of time for the company while supporting the business action. Champion and challenger models should be registered and published for future deployment.

Registering models means putting the models in a central repository. *Publishing the model* means putting the score code where the model is going to run in production. The champion and challenger models should be ready to be deployed and executed in multiple platforms.

Monitoring

An essential phase in the analytical life cycle is monitoring the model performance over time. All models are trained based on a particular data set, considering a specific time frame. Customers' behavior changes over time, and the data used to create the models are no longer the same. Because of that, the model in production was trained on a data set different from the one used to score. For this reason, the model's decisions can drift over time and the performance can decay significantly. When this happens, it is time to change the model. The champion model currently in production can be retrained. In some situations, a retrain suffices. In others, just a model retraining is not enough, and a brand-new and different model should be trained. The only way to make the company aware of this need – retrain or rebuild – is by monitoring the current models in production and evaluate when the model's performance decays.

The analytical life cycle considers multiple phases, from data preparation to retraining or rebuilding the models, passing through data exploration, modeling, deployment, and monitoring. Data scientists are involved in the initial phases of the analytical life cycle: data preparation, data exploration, and of course, modeling. They will be involved during the model assessment and deployment tasks associated with registering and publishing the champion and challenger models in different platforms or exposing them in different formats.

Model Operationalization

Many organizations in all industries struggle to operationalize their analytical models. Some research shows that only half of the analytical models developed are put in production. The other half simply does not bring any value to the organization. Other research shows that for the models that are put in

Figure 9.11: DevOps

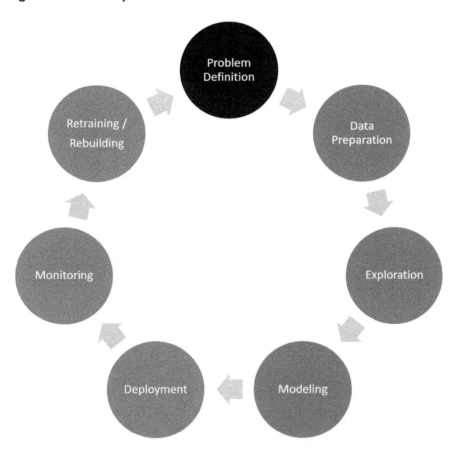

production, they take at least three months to be ready for deployment. This time and effort equals a real operational cost and means a slower time to get value from the model. In some cases, no value at all is brought to the company when the business event is time-sensitive like churn, fraud, risk, and some sales campaigns.

To overcome this challenge, many organizations are seeking a better approach to not just deploy their analytical models in an effective manner but mostly to manage and monitor them efficiently. One approach is to operationalize and integrate the tasks assigned to register, publish, deploy, monitor, and retrain the models.

All models degrade, and if they are not given regular attention, performance suffers. Models are like cars: to ensure quality performance, you need to perform regular maintenance. Model performance depends not just on model construction, but also on data quality, fine-tuning the model, regular updates to the model, and retraining the model.

Operationalizing this cycle means integrating all the steps assigned to extract value from the analytical models. This integration process is often referred to as ModelOps. ModelOps, or model operationalization, allows companies to move analytical models from the development environment, including training, validation, and testing, to production as quickly as possible while ensuring quality results. It enables companies to manage and scale models to meet demand and continuously monitor the analytical models to spot and eventually fix early signs of degradation. Speed and quality in moving models from development to production are essential to allow companies to get the most out of analytics. Monitoring and retraining are crucial to identify the right moment the models start decaying and then trying to fix them by just retraining the model. Another fix is just trying out the challenger models registered and published in the production environment. *Registering models* means putting the models in a central repository. *Publishing the model* means putting the score code where the model is run in production.

Monitor Performance

The first step in ModelOps is to monitor the performance of the models put in production. ModelOps represents a cycle involving development, testing, deployment, and monitoring. However, ModelOps is effective if it moves toward the company's main goals, which in terms of analytics are to provide a scalable approach to constantly deliver accurate analytical models in supporting business actions.

At the strategic level, organizations need to evaluate if ModelOps best practices are properly implemented, helping these companies to achieve scale and accuracy when running analytical models in production to support business actions. At the operational level, companies need to monitor the performance of all models deployed.

Very often, a dashboard is available to evaluate some fit statistics assigned to the analytical models in production. Overall misclassification or average square error for supervised models with categorical and continuous targets is an example. Some more specific metrics, such as sensitivity or positive predictive value, can also be monitored. This dashboard can also monitor some business metrics affected by the models deployed. An example is the rate of the loans approved or denied, or the rate of the customer churn.

Finally, in terms of metrics, there are some more operationalized measures to be monitored. The concepts of verification and validation are both important here. Verification confirms if a model was correctly deployed in production and performs as designed. Validation confirms if a model provides the results as expected, for example, based on the target defined to the analytical model. Both verification and validation are relevant best practices in the development and deployment of high-quality models in production.

A key component in ModelOps is indeed to monitor analytical models in production. This is to avoid the operational cost of having a useless model in production supporting a business action. All analytical models degrade, as stated before. It is just a matter of time. Some models can degrade as soon as they are deployed. For example, a fraud detection model is trained based on a particular type of fraud. As said before, many models take at least three months to reach the deployment phase. When the fraud model is deployed, the fraudster might have changed their behavior or even have changed the type of the fraud that was committed. Then, the type of the fraud the model was trained to detect no longer exists, and this model is useless.

Data Quality

Model performance can be affected by many circumstances like data quality issues, time to deployment, or simply degradation. Data quality issues are mostly associated with changes or shifts in the data that were used to train and currently used to score. Changes in the data might affect substantially the model's performance. These changes might occur in the data sources, in the input variables, or in the data transformation tasks, among others.

Time to Deployment

Time to deployment can affect the model's performance in two main situations. The first one is when the data changes significantly throughout the process from model development to model deployment, as described in the fraud detection modeling. The second one is the time to action. Some business targets are very time sensitive, and the action needs to be taken in a very specific time frame. If the action is not taken in that time frame, the target loses its meaning and the model might run ineffectively. For example, in the collections process, every step or action the company takes is associated to a time frame. If the model is deployed out of this time frame, the target is no longer valid for that model, and then the model is completely useless.

Degradation

Degradation is going to happen to any model. However, promptly capturing when this degradation happens can make the entire difference for the company in getting the most out of the analytical models deployed. It allows the company to quickly evaluate alternatives for the decayed model, either retraining or replacement. Monitoring also allows companies to improve a current model's performance. Imagine a situation where a model needs to be deployed in a specific time frame to support a campaign. The data scientist developed the model, but they believe that the model can be enhanced by using new data not considered at the beginning of the project. The data scientist can deploy the current model as they keep working on the enhancement of the new model as new data is incorporated to the data source. Once this new model is fit, it can be moved to production and run in parallel to the current model to evaluate performance in real time. If the new model provides better results, it can replace the current model easily if the ModelOps infrastructure is effectively integrated.

The great benefit of ModelOps is that it allows companies to thrive in their analytical journeys. The predictive power of these models, in combination with the availability of big data and increasing computational power, continues to be a source of competitive advantage for smart organizations, no matter the industry, no matter the size. ModelOps allows these smart companies to embrace the entire analytical life cycle in an effective manner, improving each step of the process, from model development to model deployment, passing through performance monitoring and model replacement. ModelOps allows companies to scale their analytical efforts, giving them a real competitive advantage and extracting the most value from data science initiatives.

Summary

This chapter covered important aspects of the analytical life cycle, focusing on model training, model assessment, and deployment. Model evaluation is a crucial step in the analytical life cycle because there are many perspectives on how to evaluate the best model depending on the business problem, the need for interpretability, the necessity for explicability, and the speed of training or scoring, among many others. Finally, once the best model is selected, then we come to the most important stage for the company: model deployment. Deployment is when the company gains the major benefits of using analytics in its operational process. Campaigns, decisions, prices, and bundles are based on facts, numbers, and insight from model results, no longer from guessing or feelings. The business is embracing analytics and is making data-driven decisions.

Monitoring all models in production is also crucial to support all business actions properly. Models drift over time, and customers change their behaviors, which modifies the data that describes these behaviors, which affects the models deployed. Monitoring the models running in production allows the company to identify when the model is not performing as expected and then update it. This update might be a simple retraining or an entire model development. It does not matter. The important thing is that a low-performance model needs to be replaced.

One last piece of this puzzle is model operationalization. The model might be used in diverse ways: as a report, a simple output, or in a complex application, either embedded into transactional systems or within a specific app built for it. Model operationalization is an active part of the analytical model life cycle and needs to be considered as well.

Ready to take your SAS® and JMP® skills up a notch?

Be among the first to know about new books, special events, and exclusive discounts.
support.sas.com/newbooks

Share your expertise. Write a book with SAS.
support.sas.com/publish

Continue your skills development with free online learning.
www.sas.com/free-training

 sas.com/books
for additional books and resources.

THE POWER TO KNOW®